The
# Reference Shelf®

# Global Climate Change

The Reference Shelf
Volume 85 • Number 5
H. W. Wilson
A Division of EBSCO Information Services
Ipswich, Massachusetts
2013

**GREY HOUSE PUBLISHING**

# The Reference Shelf

The books in this series contain reprints of articles, excerpts from books, addresses on current issues, and studies of social trends in the United States and other countries. There are six separately bound numbers in each volume, all of which are usually published in the same calendar year. Numbers one through five are each devoted to a single subject, providing background information and discussion from various points of view and concluding with an index and comprehensive bibliography that lists books, pamphlets, and articles on the subject. The final number of each volume is a collection of recent speeches. Books in the series may be purchased individually or on subscription.

**Library of Congress Cataloging-in-Publication Data**

Global climate change.
    pages cm. -- (The reference shelf ; volume 85, number 5)
  Includes bibliographical references and index.
  ISBN 978-0-8242-1216-2 (alk. pbk.) -- ISBN 978-0-8242-1211-7 (alk. pbk.)   1.  Global warming. 2.  Climatic changes. 3.  Global environmental change.  I. H.W. Wilson Company.
  QC981.8.G56G548 2013
  577.27--dc23

                                            2013031867

Cover: Joe Raedle/Getty Images

*The Reference Shelf*, 2013, published by Grey House Publishing, Inc., Amenia, NY, under exclusive license from EBSCO Information Services, Inc.

Printed in the United States of America

# Contents

# 4

## Wildlife and a Changing Climate

# 5

## The Oceans and Weather

# 6

## Health and Human Factors

# Preface

## The Foundations of Climate Change

Climate change is widely considered the most significant challenge facing the modern world. It is by definition a global phenomenon and, as such, has global consequences affecting all aspects of terrestrial life.

The most fundamental aspect of climate change is its basis in atmospheric behavior. While the atmosphere appears static, it is a dynamic system in which conditions in one location affect the entire mass.

The interaction between the earth and sun is the driving force behind the dynamics of the atmosphere. The sun injects enormous amounts of energy into the terrestrial atmosphere each day in the form of heat and light. The rotation of the planet about its axis and the circulation of the planet in its orbit around the sun ensure that the incoming solar energy is distributed over the entire surface of the planet, though not uniformly and not at the same time. Were this not the case, the vast majority of the planetary surface would be entirely uninhabitable and inhospitable to life. The earth's distance from the sun establishes environmental conditions that allow water to exist in three physical states—solid, liquid, and vapor. This is essential for the existence of life on the planet.

The relationship between the three forms of water and the behavior of the atmosphere is complex, involving numerous factors that combine to generate both short-term and long-term effects within the atmospheric envelope. The molecular structure of water is unique. The water molecule consists of a single oxygen atom bonded to two individual hydrogen atoms. The oxygen atom, due to the orientation of electrons about its nucleus, bonds to the two hydrogen atoms at an angle of about 109°. In addition, water is an isomolecule—its molecules do not all lie in a straight line. The oxygen atom has extra electrons that are not involved in the bonds to the hydrogen atoms. Their presence creates a region of high negative electrical character in the center of the molecule. Each of the two hydrogen atoms presents a surface of high positive electrical character on the opposite face from the oxygen atom. As a result, a strong electric dipole exists in each of the O-H bonds of the water molecule. This enables water molecules to act like magnets, so that the positive electrical charge on the hydrogen atoms is attracted to the negative electrical charge on the oxygen atoms of other water molecules. This attraction endows water with unusual physical properties that are unlike those of other molecules of a similar size and structure. Water has unusually high freezing and boiling points (0°C or 32°F and 100°C or 212°F, respectively). The bond angles and the size of the water molecule are such that several water molecules can come together to form larger and more rigid structures, as happens when liquid water freezes to form water ice. The same intermolecular attraction prevents liquid water from turning to vapor readily. This process requires a significant input of extra energy to overcome liquid water's latent heat of vaporization.

Liquid water covers over 71 percent of the earth. Intermolecular attraction connects every molecule of water on the planet to every other molecule of water. As a result, water flowing into the Arctic Ocean at the northern edges of Canada and Russia affects—and is affected by—water evaporating from the warm surface of the Indian Ocean and the Mediterranean Sea. Water that exists in gaseous form in the atmosphere and as ice and snow is also included in this relationship.

The earth's atmosphere consists primarily of the two gases, nitrogen ($N_2$) and oxygen ($O_2$). Together, nitrogen and oxygen make up about 98 percent of the atmosphere. The remaining 2 percent consists of water vapor, carbon dioxide, methane, helium, argon, and numerous other gases in small quantities. Argon makes up about 1 percent of the atmosphere, while carbon dioxide represents no more than 0.03 percent.

Solar energy entering the atmosphere interacts with atmospheric gases in different ways. Nitrogen is essentially inert, but when molecules of nitrogen are sufficiently energized, they can become ionized and react with other gas molecules. Incoming particles ejected from the surface of the sun, known as the *solar wind,* become trapped by the earth's magnetic field. The nitrogen molecules that interact with those particles absorb their energy and become ionized with much higher energy levels. The reemission of that energy as visible light produces the auroras borealis and australis (also known as the Northern and Southern Lights).

A more significant effect of solar energy in the atmosphere is what is generally termed the *greenhouse effect.* Although the phenomenon of global warming has helped give the term a negative connotation in popular culture and the media, without the greenhouse effect functioning in the atmosphere, the earth would be an uninhabitable planet of ice. The greenhouse effect occurs as sunlight reaches the earth. Incoming infrared radiation from the sun strikes the ground and is absorbed during the daytime hours. In darkness, however, this trapped heat energy is reemitted back into the atmosphere and out into space. As it passes through the atmosphere, molecules of some of the component gases—particularly, water vapor, carbon dioxide, and methane—capture some of that energy and are temporarily raised to a more energetic state. When those excited molecules release that extra energy and return to their normal energy state, much of it is returned toward the ground and laterally into the atmosphere. In this way, the atmosphere and the ground play a sort of ping-pong game with some of the energy that has arrived from the sun. The amount of energy that is retained in the atmosphere in this way is sufficient to maintain the average temperature of the planet above the freezing point of water. Without the greenhouse effect, the average temperature worldwide would be about −18°C (−0.4°F), well below the freezing point of fresh water. The planet would exist in a perpetual ice age.

Water has two other important functions in the physical regulation of the planetary atmosphere. One is in determining the planet's albedo, or the relative reflectivity of the planet's surface. Ice cover over the surface reflects incoming light away from the planet rather than allowing it to strike the terrestrial surface and be absorbed. This reflection mechanism is another factor in the balancing act that maintains a habitable terrestrial environment. Too much ice cover reflects away solar energy and can cause the average planetary temperature to fall dramatically. This is believed to be the mechanism by which several ice ages have occurred in the past, as a general cooling trend resulting from ice cover expanding beyond the point at which internal thermal sources could maintain balance. The atmosphere cooled as the process continued and more ice accumulated as fresh water from the atmosphere became trapped, disrupting the normal hydrological cycle. It is estimated that ocean water levels during the ice ages have been as much as one hundred meters lower than they are today.

On the other hand, a warming trend in the atmosphere decreases the amount of ice coverage on the planet, lowering the planetary albedo. Atmospheric warming is commonly referred to as *global warming* or *climate change.* Global warming allows more solar energy to reach the terrestrial surface and be absorbed. This additional solar energy also allows more water vapor to accumulate in the atmosphere, where it contributes significantly to the greenhouse effect, being much more effective in this process than carbon dioxide is. Ocean water levels rise as well because most of the ice that was covering the polar region has begun to melt. In past millennia, this has resulted in the formation of great inland seas.

Water's other important feature is its ability to dissolve a vast assortment of materials, even if only in exceedingly small amounts. The world's oceans are so vast and contain so much water that the amount of material dissolved in them is indeed huge. It is estimated, for example, that if all of the salt dissolved in the oceans were taken out, it would cover the dry land area of the planet to a depth of more than two meters. Some 250,000 tons of dissolved gold are also thought to be in the waters of the world's oceans, even though that element dissolves extremely poorly. Materials dissolved in a solvent act to lower, or depress, the freezing point of the solution relative to that of the pure solvent. Thus, pure water freezes at a temperature of 0°C (32°F), while a saturated solution of sodium chloride (common salt) in water freezes at a temperature of about −18°C (0°F). This freezing point depression ensures the oceans do not freeze solid, even though the water temperature may be significantly below 0°C.

Overall, the condition of the earth's atmosphere is the result of a planetary balancing act between incoming solar energy and energy emitted into space from the planet's surface. The rotation of the planet within the envelope of the atmosphere imparts lateral movement to the atmosphere, while convection cells within it impart vertical movement. Warm, humid air rising from equatorial latitudes moves toward the polar regions, becoming cool enough to condense the moisture and produce rain in the temperate latitudes. The cold, dry air that results descends again, producing the arid zones that include the Sahara, Sonora, and Gobi Deserts in the Northern Hemisphere as well as the Australian, Atacama, and Kalahari Deserts in the Southern Hemisphere. The dry winds there acquire both humidity and heat as they continue to move in polar directions, rising again in a secondary convection belt to finally come down in the Arctic and Antarctic (where some of the world's most powerful winds have been recorded) and turn back again toward the equator. Thus, the atmosphere is in a state of perpetual dynamism—a state in which many factors are involved.

The challenge for humanity is to understand this dynamism and the ways in which human activity affects the delicate balance that maintains this planet's hospitality for life. In the first two hundred years following the Industrial Revolution, human activity injected about a billion tons of carbon dioxide into the atmosphere. Based in part on analysis of gases entrapped in bubbles in the ice of Antarctic glaciers, many scientists argue that carbon emissions from human industrial activity have resulted in global warming. Others have countered that such conclusions are erroneous and argued that the amount of carbon dioxide that has been added to the atmosphere by humankind is not sufficiently significant to account for climate change. Still others argue that carbon dioxide is not the culprit and that methane and refrigerant materials known as chlorofluorocarbons (CFCs) that have been leaked into the atmosphere over several decades of common use as aerosol propellants have brought about the current climate change. Some argue that climate change is the result of naturally occurring solar and terrestrial cycles that have nothing whatever to do with human activity.

Regardless of the cause, the modern world must come to terms with the reality of climate change. Since the greater portion of the world's human population lives within sight of an ocean coastline, rising ocean water levels pose the risk that many of the world's greatest population centers will become uninhabitable. It is estimated that a rise of just one meter in the water level of the world's oceans would effectively eliminate the eastern seaboard of North America and flood much of its interior plains, while destroying much of the most densely populated regions of India and Southeast Asia. Scientists, environmental activists, and political leaders worldwide continue to debate how the modern world can turn back increases in atmospheric temperatures in order to prevent such catastrophic scenarios from occurring in the future.

# 1

# Scientific Debate on Climate Change

Hurricane Sandy is seen churning towards the east coast of the United States in this National Oceanic and Atmospheric Administration (NOAA) handout satellite image taken on October 27, 2012.

# Scientists at Odds

Science is both a human construct and a human endeavor. As such, it is rigorously subject to human nature and the subjectivity of ideas. There are few fields of scientific study in which this is more apparent than the field of climate change studies. Climate scientists have long been aware that climate change is a reality that has existed for as long as the earth has had an atmosphere and that it will continue to exist until such time as that atmosphere ceases to exist. The phenomenon of climate change, both in terms of the warming and cooling of the atmosphere, is part of an atmospheric system in perpetual dynamic equilibrium. The debate about climate change in modern science centers on what the causes of increasingly warmer atmospheric temperatures are and whether those increases can be reversed by large-scale modifications of human behavior.

## Climate Science and Human Nature

The 2006 documentary film *An Inconvenient Truth* helped launch the issue of global warming into the public consciousness. The film features a presentation by then vice president Al Gore focusing on how human industrial activity has led to an increase in atmospheric temperatures. The film became a focal point of a worldwide activism campaign aimed at raising awareness about global warming. Although the film was criticized by global warming skeptics, it was otherwise widely acclaimed, earning an Academy Award for documentary feature in 2007. That same year, Gore and the Intergovernmental Panel on Climate Change were awarded the Nobel Peace Prize.

While public awareness regarding the issue of climate change has increased since the late 2000s, the fact remains that climate science remains poorly understood. Climate scientists continue to debate both the interpretation and the validity of temperature data as they seek to identify whether a human-caused general temperature rise is the forerunner of future warming or part of a natural atmospheric cycle. While most climate scientists agree that atmospheric temperatures have increased, hyperbolic media reports warning of imminent catastrophe have helped to sensationalize the issue.

That many scientists cleave to their own ideas about climate change much more readily than accept the opposing ideas of others in the same field is a reflection of human nature. Nevertheless, as in all sciences, in climate change studies, the same data can be interpreted in different ways. This situation is made more complex when scientists from fields other than climatology add their voices to the climate science debate—a state of affairs that has been compared to getting advice about cardiology from a dentist rather than from a heart specialist, and vice versa. Within all fields of science, each individual has his or her own understanding of relevant principles. Specialization allows some scientists to know much about a particular aspect of their field but very little about everything else. This is as true of climate science as it is of chemistry, physics, and all other sciences. Within the scientific community, a democracy of popular opinion exists: an idea or theory becomes widely accepted and adopted by the public when the majority of scientists come to accept it as accurate. Those who hold differing ideas tend to be relegated to outsider status until the generally accepted opinion changes in their favor.

Within the realm of climate science, research and human nature often come into conflict. The earth's atmosphere is not an easily regulated scientific experiment, where each individual factor is carefully controlled. Measurements of any kind, such as atmospheric carbon dioxide levels and temperature, are made in the context of the local conditions that exist within the larger system. Any number of individual theories may be developed from an analysis of available data. Accordingly, each scientist that propounds a particular theory is free to deny the theories of others. The range of possible interpretations available to climatologists serves to make the debate over the causes of climate change more complicated and makes prognostications about the future of climate change and its impact on the planet more difficult.

## Is Climate Change Real?

The majority of climate scientists agree that climate change is a real phenomenon. What is at issue, however, is the extent to which climate change has occurred in the past and is occurring at present and how this phenomenon will affect the future of humankind. Measurements of the gas content of air bubbles trapped in glacial ice in various parts of the world provide some data regarding the proportions of gases in the atmosphere during different time periods. Some scientists take those analyses to be a scientific absolute and stand by the comparison with the atmospheric composition of the present time. Other scientists criticize the assumption that the gas content of trapped bubbles has not changed in the many thousands of years that have elapsed since they were formed. They deny that this assumption is valid on the understanding that certain gases can diffuse away through the surrounding ice while certain other gases cannot. Such migration, they believe, would inevitably alter the composition of the gas mixture remaining within the bubble. In addition, they argue that the sample size of such an analysis is exceedingly small relative to the planet's atmosphere.

Determining the average temperature of the planet is another area in which the data are hotly contested. The majority of scientists interpret the data as showing a steady increase in global temperatures over time, usually accompanied by assertions that the observed increase corresponds to the rate of carbon dioxide released into the atmosphere through human activities since the beginning of the first Industrial Revolution. However, other scientists criticize this analysis based on their interpretation of measured data that do not support the increasing temperature model. They also cite the known global temperature–controlling cycles of the earth's oceans. There is also research that ties observed increases in average global temperature not to the release of carbon dioxide but rather to the use of chlorofluorocarbons and shows that the average temperature has actually decreased in recent decades. There is a contingent of climate scientists who correlate global warming with solar activity and terrestrial cycles. They conclude that global temperatures fluctuate regularly according to those cycles and that the phenomena have no human cause.

## The Nature of Scientific Models

Just as the atomic theory was developed to enable the prediction of the behavior of matter, climate models have been developed with the goal of being able to predict the behavior of the earth's atmosphere. Climate scientists tend to accept the current climate change models as sufficiently predictive to make far-reaching assertions about climate and climate change, despite their models and assertions having been built upon a limited data set. The past two hundred to three hundred years, during which human activity is believed to have injected more than a billion tons of carbon dioxide into the atmosphere, represents a fraction of the atmospheric history of the planet. Some climate scientists argue that predictions made based on such a data set are inaccurate.

## The Role of Funding Sources and the Media in Scientific Debate

Much of the scientific research that occurs in the field of climate science is government funded. The use of public funds to further climate science endeavors has helped to politicize the issue. Additionally, the results of privately funded research initiatives can be influenced in such a way as to reflect the opinions of those who make said research possible. While reputable scientific institutions work to maintain objectivity in their work, there are occasions where political interests or corporate malfeasance shape the results of research work in climate science so that it supports a specific agenda. Moreover, the media plays a significant role in influencing public understanding of climate research and shaping the development of policy efforts aimed at curbing future temperature increases. Many media outlets sensationalize the issue of global warming in order to boost their audience, citing data pertaining to the issue regardless of its source or accuracy. In this way, popular opinion is manipulated by incorrect information and by data that is incomplete, outdated, or taken entirely out of context by people with little or no scientific training. Scientific research on climate change that is published in peer-reviewed journals is often a year or more out of date by the time it is published. Furthermore, the data on which such papers are based may be years old before the report is presented for peer review prior to publication.

It is often said that there are in fact three sides to every story—yours, mine, and the "truth." This is the essence of scientific debate, especially in fields involving as many random and unknown variables as climate science. Most scientists accept climate change as a reality that humankind must reckon with sooner or later. Many scientists agree that temperatures on the earth and in its atmosphere have increased as a result of carbon emissions emitted into the air as a by-product of human industrial activity. Although several contributing factors and possible caus-ative agents have been elucidated, this hypothesis remains difficult to prove con-clusively. For that reason, the scientific debate on climate change continues. There can be little doubt that the issue will remain at the forefront of public and scientific debate for years to come.

# Former Global Warming Supporter Now Shows Data That Refutes It

By Bob Adelmann
*The New American,* August 9, 2012

In his testimony August 1 before the Senate Environment and Public Works Committee, climate scientist John Christy revealed the results of his latest work showing "clear evidence...that extreme high temperatures are not increasing in frequency, but actually appear to be decreasing." Christy does his research at the University of Alabama, monitoring global temperature changes through remote satellite sensing which he developed along with a partner, Roy Spencer. For his efforts, Christy has been awarded NASA's Medal for Exceptional Scientific Achievement and the American Meteorological Society's "Special Award."

In his testimony, Christy presented evidence that current concerns about recent events—heat waves, wild fires, droughts, freak storms, and flooding—do not in fact reflect climate change (global warming) caused by humans, and concerns about human-caused global warming are in fact overblown. What Christy did was analyze the number of state record high and low temperatures, by decade, and concluded that "since 1960, there have been more all-time cold records set than hot records in each decade."

When critics noted that there are only 50 states with this data—a possibly statistically insignificant number from which to draw such conclusions—Christy then looked at "the year-by-year numbers of daily all-time record high temperatures from a set of 970 weather stations with at least 80 years of record," and found that additional data not only confirmed his original conclusion, but expanded it: There were several years with more than 6,000 record-setting highs before 1940, but no years with over 5,000 record highs after 1954.

He then went on to explain that since tracking temperature change is a long-term proposition, he created 10-year moving averages of temperatures recorded from 704 stations that had at least 100 years of data. The result? Average high temperatures have dropped significantly since the 1930s, with averages running at about half of those recorded back then.

He then answered the question he knew panelists would be asking: What about 2012? Isn't the continuing string of unseasonably hot temperatures being experienced across the country evidence of global warming?

Based on his analysis of record high temperatures in seven central US states and the West Coast, the current continuing heat wave "is smaller than previous events" going back to the 1930's. Christy then went on to answer the same question about the current drought, with the same answer: There is no discernible trend toward excessive drought conditions when compared to data going back to the year 1900.

Christy goes where the data takes him, without apparent ideological overtones or agendas. In 2003, Christy co-authored a study for the American Geophysical Union (AGU)—back when the global warming crowd was coming into full bloom—which concluded that human activities were warming Earth's climate at a faster rate than ever before.

But in a telephone interview following the publishing of that study, Christy warned about being too dogmatic:

> It is scientifically inconceivable that after changing forests into cities, turning millions of acres into farmland, putting massive quantities of soot and dust into the atmosphere and sending quantities of greenhouse gases into the air, that the natural course of climate change hasn't been increased in the past century …

[But I am] still a strong critic of scientists who make catastrophic predictions of huge increases in global temperatures and tremendous rises in sea levels.

Christy continued his research, and when the AGU issued a follow-up statement in 2007, Christy backed away from it. In an interview with *Fortune* magazine, Christy said:

> As far as the [2003 statement from] the AGU [is concerned], I thought that was a fine statement because it did not put forth a magnitude of the warming. We just said that human effects have a warming influence, and that's certainly true. There was nothing about disaster or catastrophe. In fact, I was very upset about the latest AGU statement [in 2007]. It was about as alarmist as you can get.

His work, and his commitment to take it where it led him, resulted in his writing an editorial for the *Wall Street Journal* that separated him from many of his colleagues:

> I'm sure the majority (but not all) of my IPCC colleagues cringe when I say this, but I see neither the developing catastrophe nor the smoking gun proving that human activity is to blame for most of the warming we see.

Christy was referring to the UN-established Intergovernmental Panel on Climate Change (IPCC), which is blatant in its support of the Kyoto Protocol. Its primary activity is publishing special reports on topics supportive of the implementation of that treaty. But Christy appears not to care much about what those colleagues think. In 2009, he told the US House Ways and Means Committee:

> From my analysis, the actions being considered to "stop global warming" will have an imperceptible impact on whatever the climate will do, while making energy more expensive, and thus have a negative impact on the economy as a whole. We have found

that climate models and popular surface temperature data sets [like those of the IPCC] overstate the changes in the real atmosphere and that actual changes are not alarming.

Coming from such a stalwart defender and practitioner of the scientific method which, according to the Oxford English Dictionary, is "a method or procedure that consists in [the] systematic observation, measurement, and experiment, and the formulation, testing, and modification of hypotheses," whereby such work lets reality speak for itself, Christy is sure to infuriate those whose minds are already made up about "global warming"—oops, man-made climate change—and don't want to become confused by the facts.

# 2009: Second Warmest Year on Record; End of the Warmest Decade

By Adam Voiland
*NASA Magazine,* January 1, 2010

2009 was tied for the second warmest year in the modern record, a new NASA analysis of global surface temperature shows. The analysis, conducted by NASA's Goddard Institute for Space Studies (GISS) in New York City, also shows that in the Southern Hemisphere, 2009 was the warmest year since modern records began in 1880.

Although 2008 was the coolest year of the decade—due to strong cooling of the tropical Pacific Ocean—2009 saw a return to near-record global temperatures. The past year was only a fraction of a degree cooler than 2005, the hottest year on record, and tied with a cluster of other years—1998, 2002, 2003, 2006 and 2007—as the second warmest year since recordkeeping began.

"There's always an interest in the annual temperature numbers and on a given year's ranking, but usually that misses the point," said James Hansen, the director of GISS. "There's substantial year-to-year variability of global temperature caused by the tropical El Niño-La Niña cycle. But when we average temperature over five or ten years to minimize that variability, we find that global warming is continuing unabated."

## The Hottest Decade

January 2000 to December 2009 was the warmest decade on record. Throughout the last three decades, the GISS surface temperature record shows an upward trend of about 0.2°C (0.4°F) per decade. Since 1880—the year that modern scientific instrumentation became available to monitor temperatures precisely—a clear warming trend is present, though there was a leveling off between the 1940s and 1970s.

The near-record temperatures of 2009 occurred despite an unseasonably cool December in much of North America. High air pressures in the Arctic decreased the east-west flow of the jet stream (a fast flowing air current in the troposphere), while also increasing its tendency to blow from north to south and drag cold air southward from the Arctic. This resulted in an unusual effect that caused frigid air from the Arctic to rush into North America and warmer mid-latitude air to shift toward the north.

"Of course, the contiguous 48 states cover only 1.5 percent of the world area, so the US temperature does not affect the global temperature much," said Hansen.

In total, average global temperatures have increased by about 0.8°C (1.4°F) since 1880. "That's the important number to keep in mind," said Gavin Schmidt, another GISS climate researcher. "In contrast, the difference between, say, the second and sixth warmest years is trivial since the known uncertainty—or noise—in the temperature measurement is larger than some of the differences between the warmest years."

## Decoding the Temperature Record

Climate scientists agree that rising levels of carbon dioxide and other greenhouse gases trap incoming heat near the surface of the Earth and are the key factors causing the rise in temperatures since 1880. But these gases are not the only factors that can impact global temperatures.

Three others key factors—changes in the sun's irradiance (energy), changes in sea surface temperature in the tropics, and variations in aerosol levels in the atmosphere—can also cause slight increases or decreases in the planet's temperature. Overall, the evidence suggests that these effects are not enough to account for the global warming observed since 1880.

El Niño and La Niña are prime examples of how the oceans can affect global temperatures. They describe abnormally warm or cool sea surface temperatures in the South Pacific that are caused by changing ocean currents.

Global temperatures tend to decrease in the wake of La Niña, which occurs when upwelling cold water off the coast of Peru spreads westward in the equatorial Pacific Ocean. La Niña, which moderates the impact of greenhouse-gas driven warming, lingered during the early months of 2009 and gave way to the beginning of an El Niño phase in October that's expected to continue in 2010.

An especially powerful El Niño cycle in 1998 is thought to have contributed to the unusually high temperatures that year, and Hansen's group estimates that there's a good chance 2010 will be the warmest year on record if the current El Niño persists. At most, scientists estimate that El Niño and La Niña can cause global temperatures to deviate by about 0.2°C (0.4°F).

Warmer surface temperatures also tend to occur during particularly active parts of the solar cycle, known as "solar maximums," while slightly cooler temperatures occur during lulls in activity, called "solar minimums."

A deep solar minimum has made sunspots a rarity in the last few years. Such lulls in solar activity, which can cause the total amount of energy given off by the sun to decrease by about a tenth of a percent, typically spur surface temperature to dip slightly. Overall, solar minimums and maximums are thought to produce no more than 0.1°C (0.2°F) of cooling or warming. As Hansen explained, "In 2009, it was clear that even the deepest solar minimum in the period of satellite data hasn't stopped global warming from continuing."

Small particles in the atmosphere called aerosols can also affect the climate. Volcanoes are powerful sources of sulfate aerosols that counteract global warming

> *Warmer surface temperatures also tend to occur during particularly active parts of the solar cycle, known as "solar maximums," while slightly cooler temperatures occur during lulls in activity, called "solar minimums."*

by reflecting incoming solar radiation back into space. In the past, large eruptions at Mount Pinatubo in the Philippines and El Chichón in Mexico have caused global dips in surface temperature of as much as 0.3°C (0.5°F). But volcanic eruptions in 2009 have not had a significant impact.

Meanwhile, other types of aerosols, often produced by burning fossil fuels, can change surface temperatures by either reflecting or absorbing incoming sunlight. Hansen's group estimates that aerosols probably counteract about half of the warming produced by man-made greenhouse gases, but he cautions that better measurements of these elusive particles are needed.

## Data Details

To conduct its analysis, GISS uses publicly available data from three sources: weather data from more than a thousand meteorological stations around the world; satellite observations of sea surface temperature; and Antarctic research station measurements. These three datasets are loaded into a computer program, which is available for public download from the GISS website. The program calculates trends in terms of "temperature anomalies"—not absolute temperatures, but changes up or down *relative* to the average temperature for the same month during the period of 1951–1980.

Other research groups also track global temperature trends but use different analysis techniques. The Met Office Hadley Centre, based in the United Kingdom, uses similar input measurements as GISS, for example, but it omits large areas of the Arctic and Antarctic, where monitoring stations are sparse.

In contrast, the GISS analysis extrapolates data in those regions using information from the nearest available monitoring stations, and thus has more complete coverage of the Earth's polar areas. If GISS didn't extrapolate in this manner, Schmidt explained, the computer software that performs the analysis would assume that areas without monitoring stations warm at the same rate as the global mean—an assumption that doesn't line up with the changes in Arctic sea ice that satellites have observed. Although the two methods produce slightly different results in the annual rankings, the decade-long trends in the two records are essentially identical.

"There's a contradiction between the results shown here and popular perceptions about climate trends," Hansen said. "In the last decade, global warming has not stopped."

# Global Warming: Faster Than Expected?

By John Carey
*Scientific American,* November 12, 2012

---

Scientists thought that if planetary warming could be kept below two degrees Celsius, perils such as catastrophic sea-level rise could be avoided.

Ongoing data, however, indicate that three global feedback mechanisms may be pushing the earth into a period of rapid climate change even before the two degree C "limit" is reached: meltwater altering ocean circulation, melting permafrost releasing carbon dioxide and methane, and ice disappearing worldwide.

The feedbacks could accelerate warming, alter weather by changing the jet stream, magnify insect infestations and spawn more and larger wildfires.

Over the past decade scientists thought they had figured out how to protect humanity from the worst dangers of climate change. Keeping planetary warming below two degrees Celsius (3.6 degrees Fahrenheit) would, it was thought, avoid such perils as catastrophic sea-level rise and searing droughts. Staying below two degrees C would require limiting the level of heat-trapping carbon dioxide in the atmosphere to 450 parts per million (ppm), up from today's 395 ppm and the preindustrial era's 280 ppm.

Now it appears that the assessment was too optimistic. The latest data from across the globe show that the planet is changing faster than expected. More sea ice around the Arctic Ocean is disappearing than had been forecast. Regions of permafrost across Alaska and Siberia are spewing out more methane, the potent greenhouse gas, than models had predicted. Ice shelves in West Antarctica are breaking up more quickly than once thought possible, and the glaciers they held back on adjacent land are sliding faster into the sea. Extreme weather events, such as floods and the heat wave that gripped much of the U.S. in the summer of 2012 are on the rise, too. The conclusion? "As scientists, we cannot say that if we stay below two degrees of warming everything will be fine," says Stefan Rahmstorf, a professor of physics of the oceans at the University of Potsdam in Germany.

The X factors that may be pushing the earth into an era of rapid climate change are long-hypothesized feedback loops that may be starting to kick in. Less sea ice, for example, allows the sun to warm the ocean water more, which melts even more sea ice. Greater permafrost melting puts more $CO_2$ and methane into the atmosphere, which in turn causes further permafrost melting, and so on.

The potential for faster feedbacks has turned some scientists into vocal Cassandras. Those experts are saying that even if nations do suddenly get serious about

---

reducing greenhouse gas emissions enough to stay under the 450-ppm limit, which seems increasingly unlikely, that could be too little, too late. Unless the world slashes $CO_2$ levels back to 350 ppm, "we will have started a process that is out of humanity's control," warns James E. Hansen, director of the NASA Goddard Institute for Space Studies. Sea levels might climb as much as five meters this century, he says. That would submerge coastal cities from Miami to Bangkok. Meanwhile increased heat and drought could bring massive famines. "The consequences are almost unthinkable," Hansen continues. We could be on the verge of a rapid, irreversible leap to a much warmer world.

Alarmist? Some scientists say yes. "I don't think that in the near term, catastrophic climate change is in the cards," says Ed Dlugokencky of NOAA, based on his assessment of methane levels. Glaciologist W. Tad Pfeffer of the University of Colorado at Boulder has examined ice loss around the planet and concludes that the maximum conceivable ocean rise this century is less than two meters, not five. Yet he shares Hansen's sense of urgency because even smaller changes can threaten a civilization that has known nothing but a remarkably stable climate. "The public and policy makers should understand how serious a sea-level rise of even 60 to 70 centimeters would be," Pfeffer warns. "These creeping disasters could really wipe us out."

Although scientists may not agree on the pace of climate change, the realization that specific feedback loops may be amplifying the change is causing a profound unease about the planet's future. "We have to start thinking more about the known unknowns and the unknown unknowns," explains Eelco Rohling, a professor of ocean and climate change at the University of Southampton in England. "We might not know exactly what all possible feedbacks are, but past changes demonstrate that they exist." By the time researchers do pin down the unknowns, it may be too late, worries Martin Manning, an atmospheric scientist at Victoria University of Wellington in New Zealand and a key player in the 2007 round of the Intergovernmental Panel on Climate Change (IPCC) reports: "The rate of change this century will be such that we can't wait for the science."

*"But we don't really know,"* *Alley says. "I'm still guessing that the odds are in my favor [in expecting a smaller rise], but I would hate for anyone to buy coastal property based on anything I said."*

## Hot Past Suggests Hot Future

One big reason scientists are becoming increasingly concerned about rapid climate change is improved understanding of our distant past. In the 1980s they were stunned to learn from the record written in ice cores that the planet had repeatedly experienced sudden and dramatic swings in temperature. Since then, they have put together a detailed picture of the past 800,000 years. As Hansen describes in a new analysis, there are remarkably tight correlations among temperature, $CO_2$ levels and sea levels: they all rise and fall together, almost in lockstep. The correlations do not prove that greenhouse gases caused the warming. New research by Jeremy Shakun of Harvard University and his colleagues, however, points in that direction, showing that the $CO_2$ jump preceded the temperature jump at the end of the last ice age. They conclude in a recent *Nature* paper that "warming driven by increasing $CO_2$ concentrations is an explanation for much of the temperature change." (*Scientific American* is part of Nature Publishing Group.)

Some changes in the past were incredibly rapid. Work on Red Sea sediments by Rohling shows that during the last warm period between ice ages—about 125,000 years ago—sea levels rose and fell by up to two meters within 100 years. "That's ridiculously fast," Rohling says. His analysis indicates that sea levels appear to have been more than six meters higher than they are today—in a climate much like our own. "That doesn't tell you what the future holds, but man, it gets your attention," says Richard Alley, a professor of geosciences at Pennsylvania State University.

Also surprising is how little extra energy, or "forcing," was required to trigger past swings. For instance, 55 million years ago the Arctic was a subtropical paradise, with a balmy average temperature of 23 degrees C (73 degrees F) and crocodiles lurking off Greenland. The tropics may have been too hot for most life. This warm period, dubbed the Paleocene-Eocene Thermal Maximum (PETM), apparently was sparked by a preceding bump of about two degrees C in the planet's temperature, which was already warmer than today. That warming may have caused a rapid release of methane and carbon dioxide, which led to more warming and more emissions of greenhouse gases, amplifying further warming. The eventual result: millions of years of a hothouse earth.

In the past 100 years humans have caused a warming blip of more than 0.8 degree C (1.4 degree F). And we are pouring greenhouse gases into the atmosphere 10 times faster than what occurred in the run-up to the PETM, giving the climate a mighty push. "If we spend the next 100 years burning carbon, we are going to take the same kind of leap," says Matthew Huber, a professor of earth and atmospheric sciences at Purdue University.

We are also shoving the climate harder than the known causes of various ice ages did. As Serbian astronomer Milutin Milankovic noted nearly 100 years ago, the waxing and waning of ice ages can be linked to small variations in the orbit and tilt of the earth. Over tens of thousands of years the earth's orbit changes shape, from nearly circular to mildly eccentric, because of varying pulls from other planets. These variations alter the solar energy hitting the planet's surface by an average of about 0.25 watt per square meter, Hansen says. That amount is pretty small. To cause the observed swings in climate, this forcing must have been amplified by feedbacks such as changes in sea ice and greenhouse gas emissions. In past warmings, "feedback just follows feedback, follows feedback," says Euan Nisbet, a professor of earth sciences at the Royal Holloway, University of London.

The climate forcing from human emissions of greenhouse gases is much higher—three watts per square meter and climbing. Will the climate thus leap 12 times faster? Not necessarily. "We can't relate the response from the past to the future," Rohling explains. "What we learn are the mechanisms that are in play, how they are triggered and how bad they can get."

## Troubling Feedbacks

The most rapid of these feedback mechanisms, scientists have figured out, involves ocean currents that carry heat around the globe. If a massive amount of freshwater is dumped into the northern seas—from say, collapsing glaciers or increased precipitation—warm currents can slow or stop, disrupting the engine that drives global ocean currents. That change could turn Greenland from cool to warm within a decade. "Greenland ice-core records show that shifts can occur very, very quickly, even in 10 years," says Pieter Tans, a senior scientist at the NOAA Earth System Research Laboratory.

When the freshwater mechanism became clear by the early 2000s, "a lot of us were really nervous," Alley recalls. Yet more detailed modeling showed that, although "adding freshwater is a scary thing, we're not adding it nearly fast enough" to fundamentally alter the planet's climate, he says.

A more immediately worrisome feedback that is beginning to bubble up—literally—involves permafrost. Scientists once thought that organic matter in the tundra extended only a meter deep into the frozen soil—and that it would take a long time for warming to start melting substantial amounts of it deep down. That assessment was wrong, according to new research. "Pretty much everything we've documented has been a surprise," says biologist Ted Schuur of the University of Florida.

The first surprise was that organic carbon exists up to three meters deep—so there is more of it. Plus, Siberia is dotted with giant hills of organic-rich permafrost called yedoma, formed by windblown material from China and Mongolia. Those carbon stores add up to hundreds of billions of metric tons—"roughly double the amount in the atmosphere now," Schuur says. Or as methane hunter Joe von Fischer of Colorado State University puts it: "That carbon is one of the ticking time bombs." More thawing allows more microbes to dine on the organic carbon and turn it into $CO_2$ and methane, raising temperatures and prompting more thawing.

The ticking may be speeding up. Meltwater on the permafrost surface often forms shallow lakes. Katey Walter Anthony of the University of Alaska Fairbanks has found methane bubbling up from the lake bottoms. Many researchers have also found that permafrost can crack open into mini canyons called thermokarsts, which expose much greater surface area to the air, speeding melting and the release of greenhouse gases. And recent expeditions off Spitsbergen, Norway, and Siberia have detected plumes of methane rising from the ocean floor in shallow waters.

If you extrapolate from these burps of gas to wider regions, the numbers can get big enough to jolt the climate. Yet global measurements of methane do not necessarily show a recent increase. One reason is that hotspots "are still pretty local," says the University of Alaska Fairbanks's Vladimir E. Romanovsky, who charts permafrost temperatures. Another may be that scientists have just gotten better at finding hotspots that have always existed. That is why Dlugokencky says, "I am not concerned about a rapid climate change brought about by a change in methane."

Others are not so sure, especially because there is another potentially major source of methane—tropical wetlands. If rainfall increases in the tropics, which is likely as the atmosphere warms, the wetlands will expand and become more productive, creating more anaerobic decomposition that produces methane. Expanded wetlands could release as much, or more, additional methane as that from Arctic warming. How worried should we be? "We don't know, but we'd better keep looking," Nisbet says.

## The Ice Effect

The feedback that scares many climate scientists the most is a planetary loss of ice. The dramatic shrinking of sea ice in the Arctic Ocean in recent summers, for instance, was not predicted by many climate models. "It is the big failure in modeling," Nisbet says. Ice on Greenland and along Antarctica is disappearing, too.

To figure out what is going on, scientists have been charting glaciers in Greenland by satellite and ground measurement and have been sending probes under the Antarctic ice shelves, "seeing things never seen before," says Jerry Meehl, a senior scientist at the National Center for Atmospheric Research.

On Greenland, glaciologist Sarah Das of the Woods Hole Oceanographic Institution watched as a lake of meltwater suddenly drained through a crack in the 900-meter-thick (3,000-foot- thick) ice. The torrent was powerful enough to lift the massive glacier off the underlying bedrock and increase the speed at which it was sliding into the ocean. In Alaska, Pfeffer has data showing that the huge Columbia Glacier's slide into the sea has accelerated from one meter a day to 15 to 20 meters a day.

In Antarctica and Greenland, large ice shelves that float on ocean water along the coast are collapsing—a wake-up call about how unstable they are. Warmer ocean waters are eating away at the ice shelves from below while warmer air is opening cracks from above. The ice shelves act as buttresses, holding back ice that is grounded on the ocean bottom and adjacent glaciers on land from slinking into the sea under gravity's relentless pull. Although the loss of floating ice does not raise sea levels, the submerging glaciers do. "We're now working hard to find out whether sea-level rise could be remarkably faster than expected," Alley says.

Ice loss is feared not just because of sea-level rise but also because it kicks off a powerful feedback mechanism. Ice reflects sunlight back to space. Take it away, and the much darker land and seas absorb more solar heat, melting more ice. This change in the albedo (reflectivity) of the earth's surface can explain how small forcings in the paleoclimate record could be amplified, Hansen says, "and the same will occur today."

So far only a few scientists are willing to go as far as Hansen in predicting that the oceans could rise by as much as five meters by 2100. "But we don't really know," Alley says. "I'm still guessing that the odds are in my favor [in expecting a smaller rise], but I would hate for anyone to buy coastal property based on anything I said."

## Forest for the Trees

The fluctuations in the earth's past climate make it clear that feedbacks will dramatically transform the planet now if we push hard enough. "If we burn all the carbon we have access to, we're pretty much guaranteed of having a PETM-like warming," Huber says. Good for Arctic crocodiles, perhaps, but not for humans or most ecosystems.

Yet what really keeps scientists up at night is the possibility that even if these particular feedbacks do not bring near-term threats to humanity, they could drive other mechanisms that do. A prime candidate is the planet's water, or hydrological, cycle. Each year brings additional evidence that climate change is causing more extreme weather events such as floods and droughts while fundamentally altering regional climates.

A recent analysis by Rahmstorf shows that heat waves like the one that devastated Russia in 2010 are five times more likely because of the warming that has already occurred—"a massive factor," he says. And new work pins the record-breaking warm 2011–2012 US winter (and record-breaking cold spell in Europe that same season) on the loss of Arctic sea ice. One suggested mechanism: with less sea ice, more Arctic water warms. The ocean releases that extensive heat in the autumn, altering pressure gradients in the atmosphere, which creates bigger bends in the jet stream that can get stuck in place for longer periods. Those bends can bask the US Northeast in winter warmth while locking eastern Europe in a deep freeze.

Making this story even more complex is the potential for ecological feedbacks. Warmer temperatures in the western US and Canada, for instance, have helped unleash an epidemic of mountain pine beetles. The insects have killed hundreds of thousands of hectares of trees, threatening to turn forests from carbon sinks (healthy trees absorbing $CO_2$) into carbon sources (dead trees decomposing). A hot spell in 2007 set the stage for the first fire on the North Slope's tundra in 7,000 years, accelerating permafrost melting and its carbon emissions in that area. Warming in Siberia is starting to transform vast forests of larches into spruce and fir woodlands. Larches drop their needlelike leaves in winter, thereby allowing the sun's heat to reflect off the snow cover and return to space. Spruces and firs keep their needles, absorbing the solar heat before it can reach the snow, explains ecologist Hank Shugart of the University of Virginia. Feedbacks from vegetation changes alone could give the planet a 1.5 degree C kick, he estimates: "We're playing with a loaded gun here."

Nisbet's own "nightmare scenario" starts with a blip in methane emissions and a very warm summer that leads to massive fires, pouring carbon into the atmosphere. The smoke and smog blanket Central Asia and weaken the monsoons, causing widespread crop failures in China and India. Meanwhile a large El Niño pattern of unusually warm water in the tropical Pacific brings drought to the Amazon and Indonesia. The tropical forests and peatlands also catch fire, injecting even more $CO_2$ into the atmosphere and putting the climate on the fast track to rapid warming. "It's a feasible scenario," Nisbet observes. "We may be more fragile than we think we are."

But just how powerful could the various feedback loops become? Climate models, which are good at explaining the past and present, stumble when it comes to predicting the future. "People can conceptualize these abrupt changes better than the models do," Schuur says. Even if the planet is in a tipping point now, he adds, we may not recognize it.

The unsettling conclusion for climate policy is that science does not have definitive answers. "We know the direction but not the rate," Manning says. Yet the uncertainties do not justify inaction, scientists insist. On the contrary, the uncertainties bolster the case for an immediate worldwide effort to reduce greenhouse gas emissions because they reveal how substantial the risks of rapid change really are. "What we're doing at the moment is an experiment comparable on a geological scale to the big events of the past, so we would expect the inputs to have consequences similar to those in the past," Nisbet says.

That is why Hansen cannot look at his grandchildren and not become an activist on their behalf. "It would be immoral," he says, "to leave these young people with a climate system spiraling out of control."

# The Last Great Global Warming

By Lee R. Kump
*Scientific American,* July 1, 2011

*Surprising new evidence suggests the pace of the earth's most abrupt prehistoric warm-up paled in comparison to what we face today. The episode has lessons for our future.*

Polar bears draw most visitors to Spitsbergen, the largest island in Norway's Svalbard archipelago. For me, rocks were the allure. My colleagues and I, all geologists and climate scientists, flew to this remote Arctic island in the summer of 2007 to find definitive evidence of what was then considered the most abrupt global warming episode of all time. Getting to the rocky outcrops that might entomb these clues meant a rugged, two-hour hike from our old bunkhouse in the former coal-mining village of Longyearbyen, so we set out early after a night's rest. As we trudged over slippery pockets of snow and stunted plants, I imagined a time when palm trees, ferns and alligators probably inhabited this area.

Back then, around 56 million years ago, I would have been drenched with sweat rather than fighting off a chill. Research had indicated that in the course of a few thousand years—a mere instant in geologic time—global temperatures rose five degrees Celsius, marking a planetary fever known to scientists as the Paleocene-Eocene Thermal Maximum, or PETM. Climate zones shifted toward the poles, on land and at sea, forcing plants and animals to migrate, adapt or die. Some of the deepest realms of the ocean became acidified and oxygen-starved, killing off many of the organisms living there. It took nearly 200,000 years for the earth's natural buffers to bring the fever down.

The PETM bears some striking resemblances to the human-caused climate change unfolding today. Most notably, the culprit behind it was a massive injection of heat-trapping greenhouse gases into the atmosphere and oceans, comparable in volume to what our persistent burning of fossil fuels could deliver in coming centuries. Knowledge of exactly what went on during the PETM could help us foresee what our future will be like. Until recently, though, open questions about the event have made predictions speculative at best. New answers provide sobering clarity. They suggest the consequences of the planet's last great global warming paled in comparison to what lies ahead, and they add new support for predictions that humanity will suffer if our course remains unaltered.

## Greenhouse Conspiracy

Today investigators think the PETM unfolded something like this: As is true of our current climate crisis, the PETM began, in a sense, with the burning of fossil fuels. At the time the supercontinent Pangaea was in the final stages of breaking up, and the earth's crust was ripping apart, forming the northeastern Atlantic Ocean. As a result, huge volumes of molten rock and intense heat rose up through the landmass that encompassed Europe and Greenland, baking carbon-rich sediments and perhaps even some coal and oil near the surface. The baking sediments, in turn, released large doses of two strong greenhouse gases, carbon dioxide and methane. Judging by the enormous volume of the eruptions, the volcanoes probably accounted for an initial buildup of greenhouse gases on the order of a few hundred petagrams of carbon, enough to raise global temperature by a couple of degrees. But most analyses, including ours, suggest it took something more to propel the PETM to its hottest point.

A second, more intense warming phase began when the volcano-induced heat set other types of gas release into motion. Natural stirring of the oceans ferried warmth to the cold seabed, where it apparently destabilized vast stores of frozen methane hydrate deposits buried within. As the hydrates thawed, methane gas bubbled up to the surface, adding more carbon into the atmosphere. Methane in the atmosphere traps heat much more effectively than $CO_2$ does, but it converts quickly to $CO_2$. Still, as long as the methane release continued, elevated concentrations of that gas would have persisted, strongly amplifying the greenhouse effect and the resulting temperature rise.

A cascade of other positive feedbacks probably ensued at the same time as the peak of the hydrate-induced warming, releasing yet more carbon from reservoirs on land. The drying, baking or burning of any material that is (or once was) living emits greenhouse gases. Droughts that would have resulted in many parts of the planet, including the western US and western Europe, most likely exposed forests and peat lands to desiccation and, in some cases, widespread wildfires, releasing even more $CO_2$ to the atmosphere. Fires smoldering in peat and coal seams, which have been known to last for centuries in modern times, could have kept the discharge going strong.

Thawing permafrost in polar regions probably exacerbated the situation as well. Permanently frozen ground that locks away dead plants for millions of years, permafrost is like frozen hamburger in the freezer. Put that meat on the kitchen counter, and it rots. Likewise, when permafrost defrosts, microbes consume the thawing remains, burping up lots of methane. Scientists worry that methane belches from the thawing Arctic could greatly augment today's fossil-fuel-induced warming. The potential contribution of thawing permafrost during the PETM was even more

> *The fossil record tells us that the speed of climate change has more impact on how life-forms and ecosystems fare than does the extent of the change.*

dramatic. The planet was warmer then, so even before the PETM, Antarctica lacked the ice sheets that cover the frozen land today. But that continent would still have had permafrost—all essentially "left on the counter" to thaw.

When the gas releases began, the oceans absorbed much of the $CO_2$ (and the methane later converted to $CO_2$). This natural carbon sequestration helped to offset warming at first. Eventually, though, so much of the gas seeped into the deep ocean that it created a surplus of carbonic acid, a process known as acidification. Moreover, as the deep sea warmed, its oxygen content dwindled (warmer water cannot hold as much of this life-sustaining gas as cold water can). These changes spelled disaster for certain microscopic organisms called foraminifera, which lived on the seafloor and within its sediments. The fossil record reveals their inability to cope: 30 to 50 percent of those species went extinct.

## Cork Knowledge

That a spectacular release of greenhouse gases fueled the PETM has been clear since 1990, when a pair of California-based researchers first identified the event in a multimillion-year climate record from a sediment core drilled out of the seabed near Antarctica. Less apparent were the details, including exactly how much gas was released, which gas predominated, how long the spewing lasted and what prompted it.

In the years following that discovery, myriad scientists analyzed hundreds of other deep-sea sediment cores to look for answers. As sediments are laid down slowly, layer by layer, they trap minerals—including the skeletal remains of sea life—that retain signatures of the composition of the surrounding oceans or atmosphere as well as life-forms present at the time of deposition. The mix of different forms, or isotopes, of oxygen atoms in the skeletal remains revealed the temperature of the water, for instance.

When well preserved, such cores offer a beautiful record of climate history. But many of those that included the PETM were not in good shape. Parts were missing, and those left behind had been degraded by the passage of time. Seafloor sediment is typically rich in the mineral calcium carbonate, the same chemical compound in antacid tablets. During the PETM, ocean acidification dissolved away much of the carbonate in the sediments in exactly the layers where the most extreme conditions of the PETM era should have been represented.

It is for this reason that my colleagues and I met up in Spitsbergen in 2007 with a group of researchers from England, Norway and the Netherlands, under the auspices of the Worldwide Universities Network. We had reason to believe that rocks from this part of the Arctic, composed almost entirely of mud and clay, could provide a more complete record—and finally resolve some of the unanswered questions about that ancient warming event. Actually we intended to pluck our samples from an eroded plateau, not from underneath the sea. The sediments we sought were settled into an ancient ocean basin, and tectonic forces at play since the PETM had thrust that region up above sea level, where ice age glaciers later sculpted it into Spitsbergen's spectacular range of steep mountains and wide valleys.

After that first scouting trip from Longyearbyen, while devising plans for fieldwork and rock sampling, we made a discovery that saved much heavy lifting. We learned from a forward-thinking local geologist that a Norwegian mining company he worked for had cored through sediment layers covering the PETM era years earlier. He had taken it on himself to preserve kilometers of that core on the off chance that scientists would one day find them useful. He led us to a large metal shed on the outskirts of town where the core is now housed, since cut into 1.5-meter-long cylinders stored in hundreds of flat wood boxes. Our efforts for the rest of that trip, and during a second visit in 2008, were directed at obtaining samples from selected parts of that long core.

Back in the lab, over several years, we extracted from those samples the specific chemical signatures that could tell us about the state of the earth as it passed into and out of the PETM. To understand more about the greenhouse gas content of the air, we studied the changing mix of carbon isotopes, which we gleaned mostly from traces of organic matter preserved in the clay. By making extractions and analyses for more than 200 layers of the core, we could piece together how these factors changed over time. As we suspected, the isotope signature of carbon shifted dramatically in the layers we knew to be about 56 million years old.

## Stretching Time

Our arctic cores turned out to be quite special. The first to record the full duration of the PETM warm-up and recovery, they provided a much more complete snapshot of the period when greenhouse gases were being released to the atmosphere. We suspected that the unprecedented fidelity of these climate records would ultimately provide the most definitive answers to date about the amount, source and duration of gas release. But to get those results, we had to go beyond extrapolations from the composition and concentration of materials in the cores. We asked Ying Cui, my graduate student at Pennsylvania State University, to run a sophisticated computer model that simulated the warming based on what we knew about the changes in the carbon isotope signatures from the Arctic cores and the degree of dissolution of seafloor carbonate from deep-sea cores.

Cui tried different scenarios, each one taking a month of computer time to play out the full PETM story. Some assumed greater contributions from methane hydrates, for instance; others assumed more from $CO_2$ sources. The scenario that best fit the physical evidence required the addition of between 3,000 and 10,000 petagrams of carbon into the atmosphere and ocean, more than the volcanoes or methane hydrates could provide; permafrost or peat and coal must have been involved. This estimate falls on the high side of those made previously based on isotope signatures from other cores and computer models. But what surprised us most was that this gas release was spread out over approximately 20,000 years—a time span between twice and 20 times as long as anyone has projected previously. That lengthy duration implies that the rate of injection during the PETM was less than two petagrams a year—a mere fraction of the rate at which the burning of fossil fuels is delivering greenhouse gases into the air today. Indeed, $CO_2$ concentrations are rising probably 10 times faster now than they did during the PETM.

This new realization has profound implications for the future. The fossil record tells us that the speed of climate change has more impact on how life-forms and ecosystems fare than does the extent of the change. Just as you would prefer a hug from a friend to a punch in the stomach, life responds more favorably to slow changes than to abrupt ones. Such was the case during an extreme shift to a hothouse climate during the Cretaceous period (which ended 65 million years ago, when an asteroid impact killed the dinosaurs). The total magnitude of greenhouse warming during the Cretaceous was similar to that of the PETM, but that former episode unfolded over millions, rather than thousands, of years. No notable extinctions occurred; the planet and its inhabitants had plenty of time to adjust.

For years scientists considered the PETM to be the supreme example of the opposite extreme: the fastest climate shift ever known, rivaling the gloomiest projections for the future. In that light, the PETM's outcomes did not seem so bad. Aside from the unlucky foraminifera in the deep sea, all animals and plants apparently survived the heat wave—even if they had to make some serious adaptations to do so. Some organisms shrank. In particular, mammals of the PETM are smaller than both their predecessors and descendants. They evolved this way presumably because smaller bodies are better at dissipating heat than larger ones. Burrowing insects and worms, too, dwarfed.

A great poleward migration saved other creatures. Some even thrived in their expanded territories. At sea, the dinoflagellate Apectodinium, usually a denizen of the subtropics, spread to the Arctic Ocean. On land, many animals that had been confined to the tropics made their way into North America and Europe for the first time, including turtles and hoofed mammals. In the case of mammals, this expansion opened up myriad opportunities to evolve and fill new niches, with profound implications for human beings: this grand diversification included the origin of primates.

## Too Fast?

Now that we know the pace of the PETM was moderate at worst and not really so fast, those who have invoked its rather innocuous biological consequences to justify impenitence about fossil-fuel combustion need to think again. By comparison, the climate shift currently under way is happening at breakneck speed. In a matter of decades, deforestation and the cars and coal-fired power plants of the industrial revolution have increased $CO_2$ by more than 30 percent, and we are now pumping nine petagrams of carbon into the atmosphere every year. Projections that account for population growth and increased industrialization of developing nations indicate that rate may reach 25 petagrams a year before all fossil-fuel reserves are exhausted.

Scientists and policy makers grappling with the potential effects of climate change usually focus on end products: How much ice will melt? How high will sea level rise? The new lesson from PETM research is that they should also ask: How fast will these changes occur? And will the earth's inhabitants have time to adjust?

If change occurs too fast or if barriers to migration or adaptation loom large, life loses: animals and plants go extinct, and the complexion of the world is changed for millennia.

Because we are in the early interval of the current planetary fever, it is difficult to predict what lies ahead. But already we know a few things. As summarized in recent reports from the Intergovernmental Panel on Climate Change, ecosystems have been responding sensitively to the warming. There is clear evidence of surface-water acidification and resulting stress on sea life [see "Threatening Ocean Life from the Inside Out," by Marah J. Hardt and Carl Safina; *Scientific American*, August 2010]. Species extinctions are on the rise, and shifting climate zones have already put surviving plants and animals on the move, often with the disease-bearing pests and other invasive species winning out in their new territories. Unlike those of the PETM, modern plants and animals now have roads, railways, dams, cities and towns blocking their migratory paths to more suitable climate. These days most large animals are already penned into tiny areas by surrounding habitat loss; their chances of moving to new latitudes to survive will in many cases be nil.

Furthermore, glaciers and ice sheets are melting and driving sea-level rise; coral reefs are increasingly subject to disease and heat stress; and episodes of drought and flooding are becoming more common. Indeed, shifts in rainfall patterns and rising shorelines as polar ice melts may contribute to mass human migrations on a scale never before seen. Some have already begun [see "Casualties of Climate Change," by Alex de Sherbinin, Koko Warner and Charles Ehrhart; *Scientific American*, January 2011].

Current global warming is on a path to vastly exceed the PETM, but it may not be too late to avoid the calamity that awaits us. To do so requires immediate action by all the nations of the world to reduce the buildup of atmospheric carbon dioxide—and to ensure that the Paleocene-Eocene Thermal Maximum remains the last great global warming.

# 2

# Climate Change Around the World

Ellen Creager/MCT/Landov

Grasshoppers mate on a branch in the rainforest of the Ecuadorian Amazon.

# Global Warming and Instability

Climate change, as the alternate term "global warming" implies, is a global concern. The simple reason for this is that it will affect all parts of the globe, not just coastal areas. Furthermore, the effects of climate change will be many and varied. Nearly everything on the planet, from the structure of the atmosphere to the economies of human society, will feel the influence of climate change. Some of those effects are predictable in the short term with current knowledge, but other effects will have long-term ramifications that are at present unpredictable.

Global warming is driven primarily by the amount of atmospheric "greenhouse gases," which trap heat that would otherwise be lost to space. In the earth's atmosphere, carbon dioxide, methane, water vapor, and other greenhouse gases intercept and absorb thermal energy reradiated from the ground before the heat can escape into space. Without the greenhouse effect, the earth would experience the temperature extremes between night and day that characterize other planets and celestial bodies without atmospheres, such as the moon. However, as the levels of carbon dioxide and other greenhouse gases in the earth's atmosphere increase due to anthropogenic emissions, so does the amount of thermal energy retained. The long-term effect of that energy is to increase the average surface temperature of the planet.

## The Accelerating Rate of Climate Change

A review of the geological history of the earth reveals that the planet's climate has changed numerous times in the past. Though the data are somewhat limited, they indicate clearly that past climate changes, apart from catastrophic occurrences, have taken place over vast periods of time. The geological evidence indicates changes on the order of one or two degrees Celsius over many millennia. In comparison, the rate of present-day climate change is approximately ten to fifty times faster than any period in the past sixty-five million years of geologic history. The current rate of climate change will likely outpace the rate at which species and ecosystems are able to adapt to the changing environment.

The amount of carbon dioxide and other greenhouse gases released into the earth's atmosphere has been increasing steadily over the past century, primarily due to human activities such as the burning of fossil fuels and deforestation. Despite concerns about global climate change and the widely held scientific view that it is driven by the level of greenhouses gases in the atmosphere, the amount of anthropogenic emissions of greenhouse gases has actually increased in recent years, particularly from rapid industrialization in nations such as China and Brazil. China's economic boom of the last several decades has largely been fueled by coal-burning power plants, making it the world's foremost carbon dioxide emitter. Meanwhile, deforestation may account for as much as 15 percent of all carbon dioxide emissions;

slash-and-burn clearing in the Brazilian Amazon between the late 1980s and early 2010s razed over 153,000 square miles of rainforest for cropland and pasture, releasing large quantities of stored carbon from plants and soil.

Another major factor that may be driving the rate of climate change is feedback mechanisms in global ecosystems. One example of such a feedback mechanism is the decreasing albedo, or reflectivity, of the polar regions. As global temperatures increase, vegetation will cover more area in the higher latitudes and the snow cover of the polar regions will be reduced; these environmental changes affect the albedo of these regions, causing more solar energy to be absorbed, rather than be reflected by snow or ice, and thereby leading to further increases in temperature.

To reduce the rate of climate change, environmental scientists estimate that global anthropogenic emissions of greenhouse gases must be reduced to zero by 2050. To further dampen the effects of climate change by the end of the twenty-first century, humans must achieve negative emissions, meaning a net removal of carbon dioxide from the atmosphere. As rising global temperatures are at least partially driven by anthropogenic emissions of greenhouse gases, unless there is a conscientious effort to curb such emissions, global temperatures will continue to rise unchecked, affecting everything from sea levels, weather patterns, food production, and ecosystems around the globe.

## Rising Water

Increasing temperatures will cause ocean levels to rise due to several factors, primarily from glacial melt and the thermal expansion of seawater. The geological record indicates that during a major ice age, sea levels have been lower by as much as three hundred meters than they are today, while sea levels have been much higher in globally warm times. Glaciers and the polar ice sheets naturally undergo seasonal variation, with snow and ice accumulating during the winter months as the atmosphere carries warm ocean water to the cold polar regions and melting again in the summer as atmospheric temperatures rise. However, higher-than-average temperatures have led to longer, hotter summers and shorter, warmer winters, causing a disruption to this delicate equilibrium and triggering glacial melt at unprecedented rates. Over the past century, the global mean sea level has risen by ten to twenty centimeters.

Another factor contributing to rising sea levels is the thermal expansion of water, by which the volume of water expands with increasing temperatures. Some scientific studies have projected sea level increases of up to two meters by 2100, which would have widespread catastrophic effects around the globe, including more powerful erosion along shorelines, flooding of wetlands and other important ecosystems, contamination of groundwater aquifers by seawater, inundation of major coastal cities, and higher storm surges. For residents of low-lying coastal regions and islands—such as Florida, the Netherlands, the Ganges Delta region, the Aleutian Islands, and many South Pacific island nations—the combination of rising sea levels and enhanced storm surge inundation will be disastrous. The low-lying island nation of Kiribati, located approximately halfway between Hawaii and Australia and home to more than one hundred thousand people, is expected to be completely submerged by 2070.

## Changing Weather Patterns

The increasing temperature of seawater will have other consequences as well—namely, an increase in the intensity and duration of cyclonic storms such as hurricanes and typhoons. The most prominent example of a hurricane influenced by climate change in recent years is Hurricane Sandy, whose unprecedented storm surge caused severe damage across the eastern seaboard of the northeastern United States. Hurricane Sandy was a late-season storm, powered by higher-than-average ocean temperatures. Early damage reports estimated that Sandy caused more than fifty billion dollars in property damage, making it the second-costliest hurricane to hit the United States since 1900. A study published in the *Journal of Quaternary Science* found that sea level rise accounted for an additional twenty-two inches of storm surge–related flooding during Hurricane Sandy. The extreme damage caused by Hurricane Sandy is largely attributed to factors related to climate change, such as an unusually warm mass of surface water in the Atlantic Ocean, the elevated sea level, and shifting wind current patterns responding to changing surface conditions.

The effects of global warming also appear to be linked to the El Niño–Southern Oscillation (ENSO) cycle that has affected both weather patterns and human societies globally for many thousands of years. The effects on weather systems from the ENSO cycle extend from eastern Europe to the east coast of Africa to the west coast of South America. Scientists have found that the ENSO cycle has been more active in the past twenty-five years than ever before in recorded history. While the exact mechanisms driving this weather pattern are not well understood, the increasingly frequent cycles of ENSO have led to more weather extremes, such as severe flooding and long-term drought.

Rising sea levels and increasing temperatures can only serve to intensify hurricanes, typhoons, and tropical storms, with the inevitable result that large populations will be forced to move farther inland as the seashores become increasingly unstable. Increased rainfall from the increased water burden of storms would also result in an increased incidence of floods affecting inland areas, which would have a major negative impact on croplands. Heavy precipitation in areas that are normally dry to arid has been occurring with increasing frequency and severity in North America and elsewhere in the world. In 2008, snow accumulations in China reached their highest levels in fifty years, hampering transportation and causing power outages, fatalities, and billions of dollars in damage. Under such conditions, flash floods and forced evacuations greatly disrupt the social and economic fabric of nations.

Scientists also believe that drying of the Amazon River basin associated with climate change, itself a by-product of widespread deforestation, may be responsible for the uptick in severe "once-in-a-century" droughts and subsequent fires in the Amazon seen in 2005 and 2010.

## Effects on Food Security

More severe weather patterns are expected to have significant effects on global food production, with more extensive crop damage likely occurring due to drought, flooding, and forest fires. A study from Stanford University found that current global maize and wheat production would be roughly 5 percent higher had agricultural lands not been affected by climate trends since 1980. Increasing temperatures will enable more moisture to evaporate into the atmosphere, leading to increases in precipitation. However, global warming has also altered ocean and air currents, meaning that average rates of precipitation will change worldwide, giving mid-latitude regions heavier rainfalls and making the equatorial regions increasingly arid. In addition to the effect on wildlife and local ecosystems, changing precipitation patterns will have disastrous effects on drinking water supplies, agricultural output, and global food security. As global rainfall patterns have become less stable and predictable, food prices have become increasingly volatile. Recent record-breaking droughts in the United States, Russia, and China led to significantly lower crop yields, driving up the cost of food worldwide.

The effects of rising temperatures on pests, blights, and pollinators, such as bees, are as yet unknown. Warming waters will also likely affect fish populations, especially freshwater species that cannot move to higher latitudes with increasing temperatures. Decreased salinity and increased carbon levels in the ocean will likely have unforeseen consequences on seafood. The effects of global warming on agricultural output severely jeopardize global food security, particularly in the tropical regions of the world. The strain on agriculture and water supplies will contribute to increased political and economic instability.

## The Social Future

Higher temperatures and more extreme weather patterns are expected to destabilize global food production, driving food costs up worldwide. The increasing price of food will affect everyone, but the impact will be greatest on the world's poorest people, who already struggle with food security. The likely consequence will be a widening gap between the world's rich and poor, which will have significant social, political, and economic effects. As inequality increases and resources grow scarce, violent conflict is expected to increase.

In a future in which sea levels have risen to force the evacuation of currently habitable coastal areas, obliterate some island nations, contaminate groundwater aquifers, and ruin adjacent agricultural land, population densities in inland areas must increase dramatically; correspondingly, so must the demand on freshwater resources. What kind of stresses this will place on societal structure can only be imagined, yet plans are being made at present to ensure the availability of freshwater resources in that future. Will freshwater become "the commodity of the future," controlled by major corporations, or will it still be considered one of the inalienable basic human rights? Movement in both directions is being made at present, with government agencies foreseeing the need to ensure reliable access to freshwater resources and for-profit companies anticipating highly profitable opportunities in controlling and corporatizing both existing freshwater resources and future polar meltwater.

While the economic costs of altering human behavior to curb the emissions of greenhouse gases will be great, the costs and consequences incurred by unchecked global climate change will most definitely be greater.

# A Light in the Forest

By Jeff Tollefson
*Foreign Affairs,* March/April 2013

## Brazil's Fight to Save the Amazon and Climate-Change Diplomacy

Across the world, complex social and market forces are driving the conversion of vast swaths of rain forests into pastureland, plantations, and cropland. Rain forests are disappearing in Indonesia and Madagascar and are increasingly threatened in Africa's Congo basin. But the most extreme deforestation has taken place in Brazil. Since 1988, Brazilians have cleared more than 153,000 square miles of Amazonian rain forest, an area larger than Germany. With the resulting increase in arable land, Brazil has helped feed the growing global demand for commodities, such as soybeans and beef.

But the environmental price has been steep. In addition to providing habitats for untold numbers of plant and animal species and discharging around 20 percent of the world's fresh water, the Amazon basin plays a crucial role in regulating the earth's climate, storing huge quantities of carbon dioxide that would otherwise contribute to global warming. Slashing and burning the Amazon rain forest releases the carbon locked up in plants and soils; from a climate perspective, clearing the rain forest is no different from burning fossil fuels, such as oil and gas. Recent estimates suggest that deforestation and associated activities account for 10–15 percent of global carbon dioxide emissions.

But in recent years, good news has emerged from the Amazon. Brazil has dramatically slowed the destruction of its rain forests, reducing the rate of deforestation by 83 percent since 2004, primarily by enforcing land-use regulations, creating new protected areas, and working to maintain the rule of law in the Amazon. At the same time, Brazil has become a test case for a controversial international climate-change prevention strategy known as REDD, short for "reducing emissions from deforestation and forest degradation," which places a monetary value on the carbon stored in forests. Under such a system, developed countries can pay developing countries to protect their own forests, thereby offsetting the developed countries' emissions at home. Brazil's preliminary experience with REDD suggests that, in addition to offering multiple benefits to forest dwellers (human and otherwise), the model can be cheap and fast: Brazil has done more to reduce emissions than any other country in the world in recent years, without breaking the bank.

The REDD model remains a work in progress. In Brazil and other places where elements of REDD have been applied, the funding has yet to reach many of its intended beneficiaries, and institutional reforms have been slow to develop. This has contributed to a rural backlash against the new enforcement measures in the Brazilian Amazon—a backlash that the government is still struggling to contain. But if Brazil can consolidate its early gains, build consensus around a broader vision for development, and follow through with a program to overhaul the economies of its rainforest regions, it could pave the way for a new era of environmental governance across the tropics.

For the first time, perhaps, it is possible to contemplate an end to the era of large-scale human deforestation.

## Lula Gets Tough

The deforestation crisis in Brazil ramped up in the 1960s, when the country's military rulers, seeking to address the country's poverty crisis, encouraged poor Brazilians to move into the Amazon basin with promises of free land and generous government subsidies. In response, tens of thousands of Brazilians left dry scrublands in the northeast and other poor areas for the lush Amazon basin—a mass internal migration that only increased in size throughout the 1970s and beyond.

But the government did not properly plan for the effect of a population explosion in the Amazon basin. The result was a land rush, during which short-term profiteering from slash-and-burn agriculture prevented anything resembling sustainable development. Environmental and social movements arose in response to the chaotic development, but it was not until the 1980s, when scientists began systematically tracking Amazonian deforestation using satellite imagery, that the true scale of the environmental destruction under way in the Amazon became apparent. The end of military rule in 1985 and Brazil's transition to democracy did nothing to slow the devastation; the ecological damage only worsened as road-building projects and government subsidies for agriculture fueled a real estate boom that wiped out forests and threatened traditional rubber tappers and native peoples. Meanwhile, the total population of the Amazon basin increased from around six million in 1960 to 25 million in 2010 (including some 20 million in Brazil), and agricultural production in the Amazon region ramped up as global commodity markets expanded.

Things began to change in 2003, when Luiz Inácio Lula da Silva, the newly elected Brazilian president, known as Lula, chose Marina Silva as his environment minister. A social and environmental activist turned politician, Silva hailed from the remote Amazonian state of Acre and had worked alongside Chico Mendes, a union leader and environmentalist whose murder in 1988 at the hands of a rancher drew global attention to the issue of the Amazon's preservation. With Lula's blessing, Silva immediately set about doing what no Brazilian government had previously attempted: enforcing Brazil's 1965 Forest Code, which had set forth strong protections for forests and established strict limits on how much land could be cleared. Doing so represented a major shift in domestic policy and was equally striking at the international level: Brazil chose to act at a time when most developing countries were resisting any significant steps to combat global warming absent the industrialized world's own more aggressive actions and provision of financial aid.

After peaking in 2004, when an area of rain forest roughly the size of Massachusetts was mowed down in a single year, Brazil's deforestation rate began to fall. Then, in late 2007, scientists at Brazil's National Institute for Space Research warned that the rate of deforestation had spiked once again. The increase coincided with a sudden rise in global food prices, which created an incentive for landowners in the Amazon to illegally clear more forest for pasture and crops. This suggested that the earlier decline in the rate of deforestation might have been driven by market forces as much as by government intervention, but Lula nevertheless doubled down on enforcement. The government deployed hundreds of Brazilian soldiers in early 2008 to crack down on illegal logging, issuing fines to those who broke the law and in some instances hauling lawbreakers to jail.

The following year, Brazil announced that its rate of deforestation had hit a historic low, and Lula pledged that by 2020 the country would reduce its deforestation to 20 percent of the country's long-term baseline, then defined as the average from 1996 to 2005. His plan to achieve that goal was based on one version of the REDD model, which had vaulted onto the international agenda several years earlier as scientists made advances in quantifying the impact of tropical deforestation on climate change.

## Green-Lighting REDD

Politicians and commentators usually describe global warming as a long-term threat, but scientists also worry about transgressing invisible thresholds and thus provoking potentially rapid and irreversible near-term changes in the way environmental and biological systems function. During the past decade, based in part on the results of intensive climate modeling, some scientists began to grow concerned that the Amazon could represent one of the clearest examples of such tipping points.

Think of the rain forest not as a collection of trees but as a hydrologic system, a massive machine for transporting and recycling water in which trees act as pumps, pulling water out of the ground and then injecting it, through transpiration, into the air. This process ramps up as the sun rises over the Amazon each day; as the forest heats up, evaporation increases, and trees transpire water to stay cool, simultaneously increasing the amount of water they take up through their roots. By constantly replenishing the atmosphere with water vapor, the Amazon helps create its own weather on a grand scale.

Humans interfere with this process whenever they chop down rain forests, and at some point, the system will begin to shut down. And this is not the only threat. Studies suggest that the Amazon could also be susceptible to rising temperatures and shifting rainfall patterns due to global warming. The nightmare scenario is known as "Amazon dieback," wherein the rains decrease and open savannas encroach on an ever-shrinking rain forest. The resulting loss of fresh water could be catastrophic for communities, agriculture, and hydropower systems in the Amazon, and dieback would have drastic effects on biodiversity and the global carbon dioxide cycle. The Amazon stores some 100 billion metric tons of carbon, equivalent to roughly a decade of global emissions. Converting carbon-rich rain forests into open savannas would pump massive quantities of carbon dioxide into the atmosphere, making it even harder for humans to prevent further warming.

Roughly 20 percent of the Amazon has been cleared to date, and there is already evidence that precipitation and river-discharge patterns are changing where the deforestation has been most intense, notably in the southwestern portion of the basin. And some scientists fear that the shifting climate may already be exerting an influence. In the past seven years, the Amazon has suffered two extremely severe droughts; normally, such droughts would be expected to occur perhaps once a century. One of the most comprehensive modeling studies to date, conducted in 2010 under the auspices of the World Bank, suggests that even current levels of deforestation, when combined with the impacts of increasing forest fires and global warming, are making the Amazon susceptible to dieback.

Such projections have heightened the sense of urgency in climate policy circles and helped focus attention on the REDD model. The concept has been around in some form for more than 15 years, but it was first placed on the international agenda in 2005 by the Coalition for Rainforest Nations, a group of 41 developing countries that cooperates with the UN and the World Bank on sustainability issues. At the core of the model is the belief that it is possible to calculate how much carbon is released into the atmosphere when a given chunk of forest is cut down. Fears that this would prove impossible helped keep deforestation off the agenda when climate diplomats signed the Kyoto Protocol in 1997. Scientists are steadily improving their methods for estimating how much carbon is stored in forests, however, and most experts agree that carbon dioxide can be tracked with enough accuracy to calculate baseline figures for every country.

Under various proposed versions of the REDD model, wealthy countries or businesses seeking to offset their own impact on the climate would pay tropical countries to reduce their emissions below their baseline levels. There is no consensus about the best way to design such a system of payments; since REDD was formally adopted as part of the agenda for climate negotiations at the UN Climate Change Conference in Bali, Indonesia, in 2007, dozens of countries and nongovernmental organizations have put forward a range of ideas. Most of these call for the creation of a global market that, like the European carbon-trading system, would allow industrial polluters to purchase carbon offsets generated by rain-forest preservation. Some environmentalists and social activists worry about the validity and longevity of such credits, as well as the prospect of banks and traders entering the conservation business. One fear is that "carbon cowboys," a new class of entrepreneurs specializing in the development of carbon-offset projects, would sweep through forests, trampling the rights of indigenous and poor people by taking control of their lands and walking away with the profits. This concern is valid, as there is always a danger of bad actors. But civil-society groups and governments, including Brazil's, are aware of the problem and are working on safeguards.

Brazilian officials have also expressed worries that the ability to simply purchase unlimited offsets would allow wealthy countries to delay the work that needs to be done to reduce their own emissions. An alternative backed by Brazil's climate negotiators and others would be a state-based funding system, in which money would flow from governments in the developed world to governments in the developing world, which would guarantee emissions reductions in return.

## Norwegian Wood

In 2008, Lula, perhaps hoping to preempt an interminable debate over how best to design a global REDD system, announced the establishment of the Amazon Fund, calling on wealthy countries to contribute some $21 billion to directly fund rain-forest-preservation measures. The proposal went against the market-based approach being pushed by the Coalition for Rainforest Nations. Based on a more conventional system of government donations, the Amazon Fund would allow Brazil to control the money and manage its forests as it saw fit. To the fund's backers, the resulting reductions in emissions would represent offsets of a sort.

Only one country decided to take up Lula's challenge: Norway, which stepped forward with a commitment of up to $1 billion. Coming well in advance of any formal carbon market and the international treaty that many hoped would be signed at the UN climate summit in Copenhagen in 2009, Norway's pledge was largely an altruistic vote of confidence in Brazil's approach, with donations conditioned on measurable progress. Since 2010, when the funding began, the Brazilian Development Bank, which manages the fund, has undertaken 30 projects, costing nearly $152 million. These projects include direct payments to landowners in return for preserving forests and initiatives to sort out disputes over landownership, educate farmers and ranchers about sustainability, and combat forest fires.

Although environmentalists and scientists have criticized some delays in the program, Brazil's deforestation rate has continued to plunge. Each year from 2009 to 2012, the country registered a new record low for deforestation; in 2012, only 1,798 square miles of forest were cleared. That is 76 percent below the long-term baseline, leaving Brazil just four percent shy of its Copenhagen commitment with eight years to go. Recent calculations by Brazilian scientists suggest that the cumulative release of carbon dioxide expected as a result of deforestation in the Brazilian Amazon dropped from more than 1.1 billion metric tons in 2004 to 298 million in 2011—roughly equivalent to the effect of France and the United Kingdom eliminating their combined carbon dioxide emissions for 2011.

REDD remains a distant promise for most landowners and communities, and the precipitous drop in deforestation in Brazil is more a function of broader government policy than the result of any individual project. Still, the Amazon Fund is demonstrating the promise and practicality of the REDD model. Although the actual cost of preventing emissions remains unclear, Brazil is offering donors carbon offsets at a discounted price of $5 per metric ton of carbon dioxide, intentionally underestimating how much biomass its forests contain in order to avoid arguments over the price. Of course, implementing the REDD model could prove significantly more expensive elsewhere. But the price would nonetheless be significantly cheaper than for many other methods of cutting emissions, such as capturing carbon dioxide from a coal-fired power plant and pumping it underground, which could cost upward of $100 per metric ton in the initial stages.

## Rousseff and the Ruralistas

Lula was succeeded by his protégé and former chief of staff, Dilma Rousseff, in 2011. Although environmentalists have been critical of her broader development agenda in the Amazon and beyond, Rousseff has upheld Lula's deforestation policies. And she has done so despite intense pressure from the so-called ruralista coalition of landowners and major agricultural interests, which currently exercises tremendous influence in Brasilia.

In the spring of 2012, the Brazilian Congress passed a bill that would have eviscerated the country's vaunted Forest Code by scaling back basic protections for land alongside rivers and embankments and offering outright amnesty to companies and landowners who had broken the law. Rousseff fought back, and a prolonged tussle ensued. The final result was a law that is generally more favorable to agricultural interests but that nonetheless retains minimum requirements for forest protection and recovery on private land.

More troubling than the new law itself, perhaps, is the political polarization that accompanied its passage. Brasilia now seems divided into rigid environmentalist and agricultural factions. Fierce opposition to Brazil's rain-forest-preservation efforts is sure to persist, and many observers fear that landowners, impatient with the slow pace of progress on REDD, will ultimately begin to test the limits of the newly revised Forest Code. As if on cue, last September, Brazilian scientists announced that deforestation was 220 percent higher in August than it had been in August of 2011. But it is too early to tell what this latest outbreak might mean. After all, prior spikes have incurred a government response, and each time the damage has been contained.

It is also worth noting that not only has Brazilian deforestation decreased overall, but the size of the average forest clearing has also decreased over time. The powerful landowners and corporate interests responsible for large-scale deforestation have apparently decided that they can no longer cut down rain forests with impunity. The upshot is that for the first time ever, in 2011, the amount of land cleared in the Brazilian Amazon dropped below the combined amount cleared in the surrounding Amazon countries, which make up 40 percent of the basin. In those countries, the trend is not so encouraging: deforestation in the non-Brazilian Amazon increased from an estimated annual average of 1,938 square miles in the 1990s to 2,782 square miles last year, according to an analysis published by the World Wildlife Fund.

## Missing the Forest for the Trees?

There was very little progress on REDD at the most recent UN climate summit, in Doha, Qatar, last November. Negotiators left the door open to a full suite of REDD-style models, from government-to-government financial transfers to a privatized carbon market, but failed to agree on the details. Regardless of which particular models are codified in a hypothetical future treaty on climate change, countries need to focus on making the money flow: some studies suggest that halving deforestation would cost $20–$25 billion annually by 2020. So far, governments have committed several billion dollars to forest protection through various bilateral and multilateral agreements. Through the UN, the industrialized countries have also made impressive commitments to combating climate change in the developing world, promising to contribute up to $100 billion annually by 2020, a portion of which could fund forest protection.

*After peaking in 2004, when an area of rain forest roughly the size of Massachusetts was mowed down in a single year, Brazil's deforestation rate began to fall.*

But it is not at all clear that this money will materialize, due in part to the current weakness of the global economy. And there is a limit to government largess. Advocates of rain-forest preservation are now trying to convince governments to commit money from revenue streams that do not depend on annual appropriations, which are more vulnerable to political and economic pressure. But that, too, is an uphill battle. Indeed, forest-preservation advocates cannot rely on governments alone; they will ultimately need to attract private-sector investment.

In the meantime, the fight against deforestation will rely on a patchwork of international partnerships and initiatives. Most significant, perhaps, Norway has transferred the model it developed with Brazil to Indonesia, which now ranks as the largest emitter of carbon dioxide from tropical deforestation. Just as in Brazil, the promise of REDD helped inspire some bold political commitments by Indonesian authorities, who have agreed to reduce their greenhouse gas emissions—most of which come from deforestation—by up to 41 percent by 2020 if international aid materializes. But Indonesia has neither the monitoring technology nor the institutional wherewithal of Brazil, so Norway's $1 billion commitment is aimed at helping the country build up its scientific and institutional capacity. Progress has been slow, but the advantage of a results-based approach, such as REDD, is that these initiatives cost money only if they yield positive results.

Brazil's experience offers some lessons for other tropical countries. The first is that science and technology must be the foundation of any solution. Brazil's progress has been made possible by major investments in scientific and institutional infrastructure to monitor the country's rain forests. Nations seeking to follow suit must invest in tools that will help them not only monitor their forests but also estimate just how much carbon those forests store. Working with scientists at the Carnegie Institution for Science, the governments of Colombia and Peru are deploying advanced systems for tracking deforestation from readily available satellite data. Combined with laser-based aerial technology that can map vast swaths of forest in three dimensions, these systems will be able to more accurately calculate and monitor stored carbon across an entire landscape—a feat that could allow these countries to leapfrog Brazil.

Brazil's Amazon Fund also shows that it is possible to move forward despite lingering scientific uncertainty about how to quantify the carbon stored in forests. Some critics of the REDD model have worried that it could draw attention away from the enforcement of existing forestry laws, ultimately increasing the cost of conservation and rewarding wealthy lawbreakers. But Brazil's experience shows that the two approaches can go hand in hand. Indeed, most of Brazil's progress to date has come from simply enforcing existing rules. The government has also created formal land reserves, outlawing development on nearly half its territory, and environmental groups have played a role by rallying public opinion and partnering with industry groups to improve agricultural practices. Still, enforcement can go only so far with the smaller landholders and subsistence farmers who are responsible for an increasingly large share of the remaining deforestation. Brazil must focus the Amazon Fund and other government initiatives on projects that will create more sustainable forms of agriculture for these small-scale farmers and ranchers.

The government also needs to look ahead. Cities in the Amazon are booming, and larger populations will translate into additional demands for natural resources and food. The Brazilian government has sought to increase agricultural productivity across the basin, recognizing that there is more than enough land available to expand production without clearing more of it. But Brazil should also encourage more forest recovery, which would bolster the Amazon's ability to produce rain and absorb carbon dioxide from the atmosphere. Globally, forests currently absorb roughly a quarter of the world's carbon emissions, thanks to the regrowth of forests cut down long ago in places such as the United States, and they could provide an even larger buffer going forward. Roughly 20 percent of the areas once cleared in the Amazon are already regrowing as so-called secondary forest. Scientists have calculated that if the government can increase that figure to 40 percent, the Brazilian Amazon will transition from a net source of carbon dioxide emissions to a "carbon sink" by 2015, taking in more carbon dioxide than it emits.

Deforestation is just one of many challenges buffeting the Amazon region, and improvements on this front should not obscure the ongoing problems of poverty, violence, and corruption. But at a time when expectations for progress on climate change are falling, Brazil has given the world a glimmer of hope. In many ways, the hard work is just beginning, but the results so far more than justify continuing the experiment.

# Global Warning

By Michael Le Page
*New Scientist,* November 11, 2012

*Five years ago, the last report of the Intergovernmental Panel on Climate Change painted a gloomy picture of our planet's future. As climate scientists gather evidence for the next report, due in 2014, Michael Le Page gives seven reasons why things are looking even grimmer.*

## The Arctic Is Warming Faster Than Predicted

Not so long ago, the Arctic Ocean was covered by thick ice several years old. Even at the end of summer, more than half of the sea surface was still shrouded in ice.

As the world has warmed in the past decades, the winter refreeze has stopped compensating for the summer melt. Heat-reflecting white ice has given way to heat-absorbing dark water; snow has melted ever earlier on surrounding lands; more heat-trapping moisture has entered the atmosphere; and bigger waves and storms have assailed weakening ice. Thanks to these feedback processes, the Arctic has begun to warm twice as fast as any other region on the planet.

By the late 1990s, the extent of sea ice had fallen to its lowest level for at least 1400 years. At the end of this summer, only a quarter of the Arctic Ocean was still covered in ice, a record low in modern times, and the total volume of ice was just a fifth of what it was three decades ago (see "Record Arctic ice loss" below). What's left is a thin layer that melts easily.

It wasn't supposed to happen this quickly. In 2007, when the Intergovernmental Panel on Climate Change (IPCC) issued its most recent report on the state of the planet's climate, the consensus was that the Arctic would not be ice-free in summer until the end of the century. We might have been unlucky: natural variability might have accelerated ice loss by pushing old, thick ice out of the Arctic. But climate models clearly underestimated the pace of change, too. Older models lacked important details, such as the melt ponds on the surface of sea ice that absorb more sunlight. The latest models, which include more processes, still suggest it will be several decades before the first largely ice-free summer occurs. But if current trends are a reliable guide, such summers will happen within a decade.

However long it takes, the continued ice loss will have many knock-on effects. These could include more extreme weather in the northern hemisphere, faster melting of the Greenland ice sheet and greater releases of carbon currently locked away in permafrost.

Other, even nastier surprises might also lie in store. Paul Valdes at the University of Bristol, UK, points out that relatively small changes in Earth's state—orbital changes, shifting ocean currents, and so on—have in the past produced abrupt climate changes. Some 5500 years ago, for instance, the lush savannahs and wetlands of northern Africa turned into the Sahara desert over centuries, or perhaps just decades. Older climate models produce such dramatic change only in response to big disturbances. It is not yet clear if the newer models are any better in this respect. They may be giving us a false sense of security, says Valdes.

## Extreme Weather Is Getting More Extreme

In 2010, Russia sizzled as the temperature hit nearly 40°C in several cities. In 2011, the "Groundhog Day blizzard" dumped astonishing amounts of snow across the eastern US and Canada. This year, too, has been one of extraordinary weather, from summer deluges in the UK to a storm knocking out power to millions in the US during a record-smashing heat wave—and of course, the devastation wrought by superstorm Sandy.

Such events fit a pattern. In a warming world, shifting rainfall and increased evaporation will lead to more droughts. A warmer atmosphere holds more water, making rainfall at times more intense.

It is difficult, though not impossible, to say how individual events are influenced by climate change. It is simpler to tell whether the overall numbers are increasing. Even at the time of the last IPCC report in 2007, the trends for extreme heat, droughts and intense rainfall were already clearly upward.

Not only are these trends continuing, but the weather is also becoming even more extreme than was predicted. For instance, a study this year of ocean-salinity data from between 1950 and 2000 by the Australian Commonwealth Scientific and Industrial Research Organization (CSIRO) found that the global water cycle—the rate at which water evaporates and falls as rain—has increased at double the pace projected by models that aim to simulate the global climate. Work by researchers from Taiwan and China found that the increase in rainfall intensity over the past three decades has been an entire order of magnitude greater than global climate models predict. As for extraordinary heat waves such as those in Europe in 2003 and 2010, events so far from the norm were only projected to occur towards the end of this century.

Climate scientist Jennifer Francis of Rutgers University thinks warming in the Arctic could be part of the explanation. As temperature differences between the north pole and the tropics fall, the polar jet stream, which pushes weather systems around, is meandering more and slowing down. This means weather patterns are more likely to get "stuck" in place. "Following the record-shattering loss of Arctic sea ice this summer, I would expect the jet stream to dish up a smorgasbord of extreme weather events all around the northern hemisphere," says Francis. Such a blocking pattern caused Sandy's abrupt turn into the US East Coast, perhaps implicating Arctic ice loss.

Studies of past climate suggest future weather could get even wilder. Between 5 million and 3 million years ago, for instance, the world was 2°C to 4°C warmer than today. According to a study by Kerry Emanuel of the Massachusetts Institute of Technology and his colleagues, published in 2010, rising numbers of hurricanes altered the distribution of heat in the oceans and ultimately flipped Earth's climate into a different state, with warmer tropics and far more hurricanes occurring over a much wider area than today. "I think it plausible that hurricane feedback could influence the way climate changes over the next few hundred years," says Emanuel. "I only wish we understood the problem better."

## Food Production Is Taking a Hit

Earlier this year, record harvests were predicted in the US, as farmers planted more to take advantage of rising prices. Instead, yields fell because of drought and record-breaking heat. The UK had a different problem: yields fell because of too much rain. With extreme weather hitting harvests in other areas, too, global food prices are soaring yet again.

That contrasts with the 2007 IPCC report, which predicted that if global temperatures rose 1.5°C or more above pre-industrial levels, greater warmth and higher $CO_2$ levels would increase yields, at least in temperate regions, Only warming of more than 3.5°C was expected to lead to a big drop in production.

But it seems climate change is already having an adverse effect even though the world has warmed just 0.8°C. Last year a team at Stanford University in California looked at global production of wheat, maize, rice and soybeans—crops that provide three-quarters of humanity's calories—from 1980 to 2008. Based on what we know about how temperature, rainfall and $CO_2$ levels affect growth, the analysis suggests that average yields are now more than 1 percent lower than they would have been with no warming. Without the fertilizing effect of increased $CO_2$, they would have been 3 percent lower. "The negative effects outweighed the positive ones," says lead author David Lobell. The analysis doesn't reflect the full effects of events such as floods, or the recent spate of extreme weather.

Wealthy countries should be able to compensate for these changes to some extent by altering what they grow and how they grow it, and by creating more heat-tolerant crop varieties. Indeed, they will have to: in 2008, the US National Bureau of Economic Research concluded that if the country's farmers kept trying to grow corn, soybeans and cotton in the same areas, yields would fall by three-quarters towards the end of the century.

Keeping up with the pace of change will not be easy, though, and with dry regions projected to get drier, irrigation water could run out in places. There is also no easy way to protect field crops from extreme heat or rainfall. With the weather projected to become more variable in some regions, and perhaps globally, the biggest problem for the world's farmers could be not knowing what to expect.

## Sea Levels Will Rise Faster Than Expected

The Summit weather station in Greenland sits more than 3000 meters up atop the country's vast ice sheet. The temperature there on a typical summer's day is a chilly −10°C. In July this year, however, the temperature rose above freezing. At one point, 97 percent of the sheet's surface was melting, leading to floods that washed away bridges. This was not a one-off event: bright snow is being replaced by dirty ice that absorbs more heat and melts faster. Along the coastline, the floating tongues of glaciers are breaking up. As these "dams" disappear, the rivers of ice behind them are accelerating and thinning.

Until recently, we thought it would be centuries before Greenland lost a significant amount of ice. The Antarctic ice sheet was expected to grow, with increased snowfall compensating for melting around the edges. The 2007 IPCC report assumed that the two ice sheets would contribute just 0.3 millimeters a year to sea-level rise for the next century.

Even then, many experts disputed this, and satellite measurements have since shown the two sheets are already losing enough ice to raise sea level by 1.3 millimeters a year and climbing. Recent modeling by researchers from the Potsdam Institute for Climate Impact Research in Germany, as well as studies of past climate, suggest that the planet will soon have warmed enough to melt Greenland's ice sheet entirely—if it hasn't already become warm enough. The question is how long the melting itself will take.

Most glaciologists now think that sea level will rise by at least a meter by 2100, and possibly by as much as 2 meters. That is enough to flood many low-lying cities or render them vulnerable to storm surges. James Hansen of NASA's Goddard Institute for Space Studies in New York is even more pessimistic. He argues that as the ice starts melting, positive feedbacks will kick in, accelerating the ice loss. Satellite observations do indeed show the rate of ice loss doubling every 10 years but, as Hansen himself points out, we cannot yet be certain that this is a long-term trend rather than a short-term blip.

Then there is the possibility of the west Antarctic ice sheet collapsing, as it has on many occasions over the past few million years in response to warming. Recent discoveries about the state of the ice and the nature of the underlying topography suggest that it could be more vulnerable to warm currents in the surrounding seas than previously expected.

Whether that happens or not, we should not be deceived by the small effects we are seeing now, says oceanographer Stefan Rahmstorf of the Potsdam Institute. "Sea-level rise is slow to start, but in the longer run will turn out to be one of the gravest impacts and longest legacies of the global warming we are causing now."

## Greenhouse Gas Levels Could Keep Rising Even if Our Emissions Stop

Only half of all the $CO_2$ we pump into the atmosphere stays there. The rest is absorbed by the land and oceans. But as the world warms, they will be able to take up less. Eventually, they will begin to emit $CO_2$.

The 2007 IPCC report included projections of increasing carbon feedbacks from seas, soils and changing vegetation, but no model included the possibility that carbon locked away in permafrost and in methane hydrates in the seabed might also be released. This is thought to have happened during past warming episodes, such as the Paleocene-Eocene Thermal Maximum 55 million years ago.

This year, researchers from the University of Victoria in British Columbia, Canada, made one of the first attempts to account for permafrost emissions and concluded that they will lead to an extra warming of around 0.25°C, and possibly 1°C, by 2100. The study assumed permafrost deeper than 3.5 meters would remain intact, and did not include erosion of coastal permafrost. It also ignored the fact that some carbon will be released in the form of the more potent greenhouse gas methane. In other words, it is likely to be an underestimate.

Surprisingly, if we emit huge amounts of $CO_2$, such feedbacks actually make less difference: the more $CO_2$ there is in the atmosphere, the less warming effect adding more has. The really worrying finding is that under certain circumstances carbon feedbacks could lead to a self-perpetuating cycle. Even if all human emissions stopped, $CO_2$ levels would continue to rise as permafrost melted, leading to further warming and carbon releases.

Simulations by Paul Higgins of the American Meteorological Society point to a similar conclusion. He says recent projections of carbon feedbacks may be overly optimistic about how much plant growth is boosted by increasing $CO_2$ levels, and how fast plants and trees colonize new areas as conditions become favorable. "Even in the lowest emissions scenario, we could end up with atmospheric $CO_2$ concentrations consistent with the worst-case emission scenario," says Higgins.

## We're Emitting More Than Ever

If we stopped pumping more C02 into the atmosphere now, we'd have a very good chance of avoiding a big hike in temperature. But there is no sign of that happening. Annual emissions fell only slightly after 2008—the biggest financial crisis since the Great Depression—and are now climbing more rapidly than ever (see "No let up" below). So far they are near the top of the IPCC's worst-case emissions scenario. "Our emissions are not slowing," says Paul Valdes of the University of Bristol, UK. 'That's the most scary aspect of our future."

The only international agreement to limit greenhouse-gas emissions, the Kyoto protocol, excluded developing countries and involved only minor cuts. The US never signed up and Canada has withdrawn. Hopes for a more effective and inclusive agreement have faded.

Meanwhile China, now the world's biggest $CO_2$ emitter, is investing heavily in renewable energy, but its rapid growth means its emissions will still soar. Some smaller countries have unilaterally promised big cuts, but few are investing to deliver them. On the contrary, most continue to subsidize fossil fuels and to build coal or gas-fired power stations, committing themselves to decades of continued emissions. The move away from nuclear power after the meltdowns at Fukushima

> *Until recently, we thought it would be centuries before Greenland lost a significant amount of ice. The Antarctic ice sheet was expected to grow, with increased snowfall compensating for melting around the edges.*

has made matters worse: even Germany, which was leading the way on expansion of renewable energy, is now planning to build more coal-fired power plants. "We are going to do virtually nothing," says Kevin Anderson of the Tyndall Centre for Climate Change Research in Manchester, UK. "Every year, it looks worse than previously."

So we are on a path that the 2007 IPCC report concluded would most probably lead to a 4°C rise in temperature by 2100—way above the 2°C level it was declared we should avoid at all costs. But this worst-case scenario was not simulated with the most advanced models when the 2007 report was being prepared because of limited time and computing power. That has since been done, and the "best estimates" are now between 5°C and 6°C by 2100, with roughly a 10 percent chance of a rise of 7°C. This means many of us are likely to live long enough to experience severe global warming. We are on track for 4°C by the 2070s, or the 2060s if carbon feedbacks (→ 5) are high. Far from being alarmist, says Anderson, most scientists have underplayed the significance of the emissions story to make their message politically more acceptable.

## Heat Stress Means Big Trouble

In August 2003, Europe was hit by an extraordinary heat wave. In parts of France, the temperature hit 40°C for seven days in a row. So many people died that a refrigerated warehouse near Paris was co-opted to store bodies. A study in 2008 concluded that the total death toll was around 70,000. Most of the victims were elderly or ill, but not all.

Heat has more subtle effects, too. The productivity of people who work in non-air-conditioned environments falls roughly 2 percent for every 1°C rise above their comfort zone. If it gets too hot, people begin to suffer from exhaustion, heatstroke and kidney failure.

Recent studies suggest the effects of climate change on human health and economic output have been underestimated. "I suspect heat stress will prove the single worst aspect of climate change," says Steven Sherwood, an atmospheric scientist at the University of New South Wales in Sydney, Australia.

What matters is not so much the air temperature but the temperature of our skin: sweating cools our skin, but is less effective in humid conditions. The combined effect of heat and humidity can be gauged by the wet-bulb temperature of a "sweating" thermometer—a thermometer wrapped in a damp cloth.

Currently, the maximum wet-bulb temperatures reached anywhere on the planet do not exceed 31°C, but we do not expect that to remain so. "All our models show a strong increase in wet-bulb temperatures, and as a result higher heat stress and health impact," says Erich Fischer at the Institute for Atmospheric and Climate Science in Zurich, Switzerland.

It is very difficult to put precise numbers on the effects because if we are healthy we don't just sit there sweating when it gets hot. We seek out cool spots, or install air conditioning, and so on. Killer heat in one country is no problem in another where people and infrastructure are adapted to it.

But there is an absolute limit. We cannot survive wet-bulb temperatures of 35°C or more for long, even standing naked in front of a fan. And a 2010 study by Sherwood and his colleague Matthew Huber concluded that if the world warms by 7°C, parts of the world will start to exceed this limit occasionally. Eventually, vast swathes of Africa, Australia, China, Brazil, India and the US will become uninhabitable for at least part of the year.

Something similar may have happened before—a mass extinction 250 million years ago is now blamed on temperatures rising too high for most animals to survive. "It looks like if we fully 'develop' all of the world's coal, tar sands, shales and other fossil fuels we run a high risk of ending up in a few generations with a largely unlivable planet," says Sherwood.

## Record Arctic ice loss

- The area of Arctic sea ice has been dropping over the past three decades, satellite measurements reveal. It reached a new low this summer.

## No let up

- Despite a dip after the 2008 financial crisis, global $CO_2$ emissions are now rising faster than ever—and closely tracking the IPCC's "worst-case" scenario. The result will be dangerous warming well before the end of the century.

- In 2012, droughts led to failed harvests in several regions including the US and Niger in Africa.

- Extreme weather across the northern hemisphere is just one consequence of Arctic warming.

## Losing its sparkle

- Greenland's ice sheet is absorbing ever more heat as dirty ice replaces bright, reflective white snow.

- Reflectivity in July 2012 compared with July average from 2000–2011

- Building more coal or gas-fired power stations puts us on the road to decades of continued emissions.

- Heat stress could make areas such as Australia's interior uninhabitable.

# Commodifying Water in Times of Global Warming

By Astrid Bredholt Stensrud
*NACLA Report on the Americas,* Spring 2011

In March 2009, Peru passed a new water resources law under President Alan García. The law was created in response to the country's growing water problem related to the threat of climate change and melting glaciers, urban population growth, pressures on limited resources, and the increasing presence of the mining industry. In the words of former president García, "The law should bring modernity to the use of water in our fatherland, modernity in the daily use of water in the households. We should all prepare ourselves to face a difficult future of the water."[1]

The idea of modernity is deeply embedded in the neoliberal project of creating a free market and in the ideal of progress, and from the government's point of view, the rural indigenous peoples in the Andes should be included in this modernity as a way out of "backwardness" and as a solution to poverty. One way of doing this is to change the campesinos' water practices and introduce new technologies for efficient irrigation. This is not a straightforward process, however, due to the steep highland landscape and campesinos' lack of financial means. Although this is part of a World Bank–funded development program, the implementation implies a lot of investment of money and time from the campesinos.

The questions that people all over Peru are asking today concern how this situation should be dealt with, and most importantly: Who is responsible for dealing with the effects of climate change? What are the consequences of these discourses of "modernity" emphasized by the central government? Who will suffer from the future water scarcity? How can it be alleviated, and who will pay the price? Indigenous campesinos in the Peruvian Andes are the ones who are most acutely experiencing the effects of global warming and who fear the threat of a waterless future. In a survey I conducted in the southern Peruvian Andes in 2011, respondents often said, "The water will decrease with global warming, and we will lose our harvests and animals," or "With time there will be no water. We will have to buy water in bottles."

Like the rest of the Global South, Peru contributes very little of the world's carbon-dioxide emissions. In a 2008 world ranking, the country came in at 143 out of 215, with 0.38 metric tons of carbon per capita. Nevertheless, global warming is producing strong observable effects on temperature, precipitation, seasonality, glacier retreat, and water supply in the Andes. After the new law, the conflicts over water have still been escalating in numbers and intensity: In 2010, the National Water Authority identified 244 social conflicts related to water resources, 22 of which were in a "critical state."[2] Most of the conflicts are related to mining activities, like the infamous Conga gold-mining project in Cajamarca in the northern Peruvian Andes, where the population opposes the project because they fear the mine will contaminate their water resources. President Ollanta Humala was elected in 2011 based on his promises to work for social inclusion and to counter the power of multinational mining companies. However, the challenge of balancing environmental concerns with hopes of economic progress has proved difficult. In Cajamarca, he explicitly stated that he favored water over gold, yet the Conga conflict continues. During the first 18 months of Humala's presidency, 15 civilians died during protests.

Two years after the new water law was passed, on March 22, 2011, the global World Water Day was celebrated in Chivay, a small town and the capital of the province of Caylloma in the southern Peruvian Andes. Delegations from the local communities in the province marched around the plaza with painted banners and shouted, "Water is life/Long live the water!" (¡El agua es vida/Que viva el agua!) and "The water should not be sold/The water should be defended!" (El agua no se vende/El agua se defiende!). Farmers and herders had traveled from all over the province to participate in the march and in the following public assembly to express their concern about the climate changes in their communities, such as melting glaciers, heavy rain, floods, irregular frosts, longer drought periods, and scarcer water supplies. A few days earlier, after a month of torrential rainfall and heavy snow in the highlands, as well as unexpected frost in the valley, the political authorities in the province of Caylloma declared a state of emergency in the province. Seventy percent of the population had been affected by the extreme weather; 25,000 young alpacas were killed, 400 hectares of cultivated land were lost, and several irrigation canals were destroyed. The mayors of Caylloma's 20 subdistricts estimated the losses to exceed $1.5 million.[3] Despite these damages caused by excessive amounts of water, the general experience in the province was that of decreasing water supplies and a fear of a waterless future.

The people gathering in Chivay on World Water Day in 2011 agreed that global warming was to blame for their increasing environmental problems. However, the question they had come to discuss was, "Which actions could be taken locally to mitigate the effects of climate change?" Furthermore, who is responsible and who should pay the price? While most of the responsibility for producing carbon emissions and global warming lies with the Global North, it is the Global South that suffers the consequences. Moreover, the groups of people who are most severely affected are already vulnerable due to poverty and unequal access to resources. A peasant family that depends on small-scale subsistence farming will have more difficulty surviving if they lose 80 percent of their harvests due to drought, flood, or frost than a large-scale agriculture company. Poor farmers can seldom afford to improve their means of production, they cannot afford insurance, and they are more dependent on government relief in case of a disaster. This raises important questions concerning responsibilities and social inequalities. Shifting responsibility from the collective to individuals has always been part of the free-market ideology. Today there is also a tendency for governments, international organizations, and NGOs to shift much of the responsibility of climate-change mitigation to individuals while often allowing corporations to continue with "business as usual."

However, the danger of this individualization of responsibility is that corporations can get away with polluting water supplies and destroying the livelihoods of peasant farmers, who then have to shoulder the burden. Most of those who live in the areas most vulnerable to climate change do not have the resources and financial means to implement the necessary projects. In this way, social inequalities will increase, both within countries and on a global scale.

Therefore, people in poor and vulnerable regions of the Peruvian Andes are now struggling to be heard and asking the government, international organizations, and multinational mining companies operating in the area to contribute financially to maintaining the headwater environment. They are specifically asking that tree-planting and micro-dam projects be supported in order to improve the soils and retain the rainwater that falls heavily in the short rainy season. The Andean farmers and herders' experience of living in a vulnerable environment, and at the same time seeing that mines or hydroelectric companies farther down the watershed use water to make money, lead to feelings of injustice and claims for collective responsibilities.

The Intergovernmental Panel on Climate Change (IPCC) states that freshwater-related issues play a pivotal role among the key regional vulnerabilities in Latin America.[4] Reports from climatological and hydrological measurements indicate that the near-surface air temperature in the tropical Andes has significantly increased over the last 70 years, and most station records in southern Peru indicate a precipitation decrease after 1950.[5] This means that rain-fed agriculture is becoming more difficult and that the glaciers are shrinking. Peru contains 70 percent of the world's tropical mountain glaciers, which are natural containers of freshwater that comes as precipitation during the rainy season.

Today, these glaciers are the most visible indicators of climate change, due to their sensitivity to increased temperatures. Glacier scientists have been monitoring the retreat and melting of these glaciers since the 1970s.[6] According to the IPCC, the total glacier area in Peru has decreased 22 percent during the last 35 years, and glacier surface from very small glaciers has fallen 80 percent in the last 30 years. It has been calculated that by 2025, all the glaciers situated 5,500 meters above sea level and under will disappear.[7] The indigenous people who reside between 3,000 and 5,000 meters above sea level live primarily from small-scale agriculture and pastoralism, and during the dry season almost all the water that people and animals use throughout the Andes is derived from the glaciers in the mountain's high peaks.[8]

Peru is vulnerable to these climate changes primarily because of the unequal distribution of resources, money, people, and water. Fifty-three percent of the population lives in poverty, and massive migration to cities in the coastal desert has been going on since the 1970s. When peasant farmers lack the economic means to improve their production under declining environmental conditions, they tend to migrate to the urbanized and industrial coast, where they live in poor neighborhoods in the desert. In Peru, 70 percent of the country's 28 million people live along the Pacific coast yet have access to only 1.8 percent of the total water resources in the country.[9] This makes water management a crucial and highly political issue.

Peru's national economy is one of the fastest growing in Latin America, in great part due to the mining industry, yet large parts of the population, especially indigenous people in the Andean highlands and in the Amazon, are still excluded from this growth and find themselves increasingly vulnerable in terms of global warming and scarcity of clean water. Poor Peruvians fear not only the effects of global warming but also the residue of more than 30 years of neoliberal globalization, such as the privatization of natural-resource extraction, the intrusion of multinational mining companies, and the tendency toward the commodification of water, particularly in moments of scarcity.

Peru's new Law on Water Resources of 2009 affirms that water is the public property of the state, but nevertheless adapts to neoliberal policies of deregulation and privatization. In 1969, the reformist Velasco government nationalized water in Peru, enacting a water law that acknowledged all water as patrimony of the state. From 1993 to 2000, the neoliberal Fujimori government attempted to privatize and create markets for water, but failed due to strong opposition from the popular social movements. The García government continued the neoliberal agenda, and in 2007, the water law was adapted to the free trade agreement with the United States. In March 2008, a legislative decree was published, promoting private investment in irrigation, and one year later, García succeeded in passing the new water resources law.

Although all forms of water are still acknowledged as the property of the state, the new law is quite ambiguous, having been modified to encompass diverging interests concerning public and private investments and responsibilities. Although water cannot be privately owned, bought, or sold, all water users have to pay different tariffs and license fees for the right to use it. Farmers pay tariffs for irrigation water and infrastructure to the regional water users' organizations, and they pay license fees to the state for the right to use a certain amount of water. If they live in a town with drinking-water facilities, they pay another tariff for this. The water users' organizations that charge the irrigation water tariffs and operate the irrigation infrastructure are encouraged to develop a "modern optic and business bias" in order to continue operating.[10] By opening up space for private investments, the law also allows private companies to operate hydraulic infrastructure and hence also makes it possible to replace the water users' own organizations.

In 2011, the Subsectoral Irrigation Program (PSI), financed by the World Bank, started to operate in Caylloma province. The aim of PSI is to achieve more efficient water use by encouraging

*These farmers make their living in an environment that has always been arid and dependent on irrigation, and they are therefore extremely vulnerable to changes in weather and climate.*

farmers to transition from traditional irrigation methods based on gravity and fur-rows to drip- and sprinkler-irrigation systems. These new technologies are embed-ded in discourses about modernity, rationality, and efficiency. However, the drip and sprinkler systems have been criticized for not being easily adaptable to the high-lands, where the terrain is steep with tiny fields and terraces. Moreover, to obtain these high-tech systems, the local communities need to finance 10 percent of the installations, which in fact amounts to very high costs for poor farmers.

This means that, once again, the peasant farmers in the Andes will lose in the modernization schemes of international-development programs. These farmers make their living in an environment that has always been arid and dependent on irrigation, and they are therefore extremely vulnerable to changes in weather and climate. Among the small-scale farmers and herders in Caylloma, the effects of global climate change are being perceived as the loss of stability; the known sea-sonal cycle of rain, frost, heat, and drought has changed, and heavy precipitation in short periods has caused erosion, landslides, and the death of animals. Most farmers in Chivay worried for future harvests since they had experienced economic loss when the premature frosts in February 2011 destroyed the crops of potatoes, peas, beans, and maize. A deep concern is the ever-decreasing water supply due to melting glaciers and drying springs. The farmer Pablo Tejada in Chivay told me, "There is not enough water. The springs are not like before. They have dried up. In my father's land, there used to be two springs that gave water, and there was a little garden there. Now they are dry." These changes are experienced locally, and at the same time the new sense of unpredictability and water insecurity are being explained by global climate conditions. Since the melting glaciers, irregular rain, and drying springs are seen as effects of global warming, water is now articulated as a finite resource, and the threat of a water crisis is increasingly being perceived as irreversible.

After the march on World Water Day in Chivay, the Water Users' Organization of Colca Valley organized a public assembly to discuss the consequences of climate change in the province and to propose solutions for the future. Representatives from communities in the highlands, mid-valley, and low valley presented their tes-timonies about how the mountains have lost their snow, the springs have dried, the water level in the rivers has decreased, and the soil has eroded due to heavy rain and lost its capacity to store rainwater.

The solutions promoted by the authorities focused on how local communities and individual farmers could take care of the headwaters and use the available water more efficiently and protect their right by complying with the required payments. The president of the water users' organization emphasized the obligation to pay the water tariff to have the right of reclaiming the use of water, and the gobernador, as representative of the government, focused on the payments of licenses to the state that would ensure the farmers their right to use water, "almost like proprietors." The representative from the Ministry of Agriculture said that "the solution is the adequate use of the water" by using new techniques to avoid wasting the water. These ideas come from the World Bank's Water and Sanitation Program, which aims to teach citizens and households that they should be more efficient in their water use and avoid wasting it in daily practices. While individual families certainly can gain by saving water, this is not a long-term solution. The problem is that "efficient use" requires technology that is hardly available for most farmers. It also does not address the fundamental problems of industrial pollution and poverty. Hence it will not reduce the continuing problem of melting glaciers and water scarcity. Short-term Band-Aid approaches and quick-fix solutions like these are inspired by neoliberal thought, which reduces collective solidarity to private moral responsibility and glosses over the underlying structural causes.

The representatives from the communities, on the other hand, focused on the importance of projects of planting trees and constructing micro-dams, which together are called "sowing and harvesting water." Although these projects will not stop global warming, they are ways of taking care of the environment in the headwaters. Planting trees prevents erosion and enables the soil to store water. The micro-dams collect water when the heavy rains come in January and February and can distribute the water in irrigation canals throughout the year. These projects were initially started by NGOs in the 1990s to improve pastures and the productivity of livestock in the headwaters, but have in later years been adopted by highland municipalities and are today seen as solutions to save water in times of climate change. The micro-dams are built in the headwaters, above 4,000 meters, which is home to the poorest people in the Andes, namely, the herders of alpacas and llamas. The water, which is "born" in their territories, flows down to the coast, where others use it as an economic resource.

The highlands of Caylloma are part of the headwaters for two watersheds: the Camaná-Majes-Colca watershed, where the water is channeled to the export agriculture in the pampa of Majes, and the Quilca-Chili watershed, where the water is used for hydropower and mining, as well as urban consumption in the city of Arequipa. These water connections raise questions about responsibility: Who is responsible for the maintenance of the headwaters—the local herders and farmers, the local or regional authorities, the government, the lowland farmers, the hydroelectric plant, or the mining companies? Therefore, on World Water Day in Chivay, the community representatives addressed the crucial question of who would finance the projects of sowing and harvesting water.

People in the headwaters see it as unjust that they should pay for projects to mitigate the effects of global warming and maintain the headwater environment, when private companies make profit on the same water further down the watershed. Therefore, the political leaders in Caylloma province are starting to formulate new demands; they claim that a multinationally owned copper mine and a state electric company should pay a canon hídrico—a social water payment—as compensation to the province, since they use water that is "born" in their poor highlands. The mayor of Caylloma province often says in his popular speeches that water is the wealth of Caylloma and that they cannot sit and watch others getting rich using this water and not compensating them. This demand is based on the valuation of water, as formulated in the water law. This strategy thus builds on the logic of commodifying water, and instead of opposing it, they make the principle of valuation their own by turning it to their advantage. They acknowledge that water has value and reclaim their part of it. Moreover, the demand for a canon hídrico is based on a notion of justice, fairness, and reciprocity: Those who benefit from the water as a resource to make a profit should contribute accordingly to the sowing and harvesting of water in order to sustainably manage the headwaters, a goal that would in turn benefit all the water users along the watershed. Hence, supporters of the canon hídrico claim that since the consequences of climate change are unevenly distributed, the responsibility of mitigating these consequences have to be equally distributed, not unlike the discussions about "climate debt" at a global level.

When I interviewed the mayor of Caylloma province, Elmer Cÿceres Llica, he said: "Those on the coast take the water for free. We send water and send water, and the coast does not even worry if the water is drying in these parts, if there are no trees, or if there are filtrations, or if a mining company enters. The poorest areas of the Peruvian Andes are those that provide water to the coast. [This should be a] part of the reciprocity that we as Andeans manage: Well, I give you water, so you should give me something back. We can restore [agricultural] terraces and build dams. But the idea is that we sow water with that money. We will sow, for example, native plants around the water sources. In other words, this is all work to preserve and harvest the water."

In a vertical landscape, where the lower parts depend on the actions in the higher parts, water management necessitates cooperation. Local environments and social groups are unevenly affected by global warming, and when responsibilities of climate-change mitigation are given to poor communities and depend on private initiatives, the environmental and social inequalities will increase. Climate change in the Andes is a kind of chronic disaster that creates winners and losers and thus power struggles within a water regime influenced by neoliberal thought and individualized responsibilities. The danger of individualized responsibility is that it allows for some to profit from investments and to shed responsibility, while the majority of those already suffering the consequences of climate change will be dependent on the goodwill of private NGOs to survive. This means that, once again, the poor and indigenous people in the Andes will lose when water is valued as a resource. Their struggle for collective responsibilities in water management is thus an attempt to take control of an uncertain future.

## Notes

1. Autoridad Nacional del Agua (ANA), Ley de Recursos Hídricos y su Reglamento, Ley no. 29338 (Lima: Ministerio de Agricultura, 2010).
2. Grupo Agua, Radio RPP, 2010, available at radio, rpp.com.pe.
3. "Colca en emergencia por lluvias y nevada," Perú 27 (Arequipa), March 18, 2011.
4. Bryson Bates, Zbigniew W. Kundzewicz, Shaohong Wu, and Jean Palutikof, eds, "Climate Change and Water," technical paper of the Intergovernmental Panel on Climate Change, Geneva, 2008, available at ipcc.ch.

5.  Mathias Vuille et al., "Climate Change and Tropical Andean Glaciers: Past, Present and Future," *Earth-Science Reviews* 89, no. 3–4 (2008): 79–96.

6.  Mark Carey, "The Politics of Place: Inhabiting and Defending Glacier Hazard Zones in Peru's Cordillera Blanca," in *Darkening Peaks: Glacier Retreat, Science, and Society,* ed. Ben Orlove, Ellen Wiegandt, and Brian H. Luckman (Berkeley: University of California Press, 2008): 229–40.

7.  María Teresa. Oré et al., El agua, ante nuevos desafíos: adores e iniciativas en Ecuador, Perú y Bolivia (Lima: Oxfam International, 2009), 56.

8.  Inge Bolin, "The Glaciers of the Andes Are Melting: Indigenous and Anthropological Knowledge Merge in Restoring Water Resources," in *Anthropology and Climate Change: From Encounters to Actions,* ed. Susan A. Crate and Mark Nuttall (Walnut Creek, CA: Left Coast Press, 2009), 228–39.

9.  CONAM, "National Communication of Peru to the United Nations Climate Change Convention," National Council on the Environment, Lima, 2001.

10. Junta Nacional de Usuarios de los Distritos de Riego del Perú (JNUDRP) and Proyecto Subsectorial del Irrigation (PSI), *Programa de fortalecimiento de organizaciones de usuarios. Legislacion para juntas de usuarios y comisiones de regantes,* ed. Ing. Javier Cupe Burga, 5th ed. (Lima: Rolling Impresores S.A., 2001).

# China Confronts Global Warming Dilemma

By Christina Larson
*Christian Science Monitor,* November 12, 2009

China awoke to climate change with a storm. It was late January 2008, a time when people across the country were busily gathering recipes, stocking fireworks, and preparing to welcome relatives to celebrate the Lunar New Year. But suddenly, severe ice storms brought much of the nation to a standstill. For two weeks, fierce winds, sleet, and snow downed power lines, shuttered businesses, and razed more than 200,000 homes across southern and central China.

Hundreds of thousands of travelers who had been headed home to see families were stranded on icy rail platforms. Cities struggled to provide power and water to residents, and snow blanketed the Taklamakan desert. Even the bright lights of Shanghai briefly went dark. All told, more than 100 people were killed.

China's worst storm in decades was, according to United Nations scientists, an illustration of what a changing climate may herald for the future. As such, it was a tipping point in the country's environmental awareness.

"For the ordinary people," says Hu Kanping, editor of the Environmental Protection Journal, "it was a historical moment for them to know what is climate change."

In an editorial comparing the storm to Hurricane Katrina, the influential Chinese business magazine *Caijing* wrote: "This painful experience ought to set us thinking about how we can better pay Nature the respect she deserves and should make us listen more attentively to what science tells us about how climate change leads to natural disaster."

The storm was also a moment for China's leaders to consider the consequences of the extraordinary growth of the country's economy, which has expanded by 9.7 percent annually over the past 30 years.

Now the world's top global exporter of manufactured goods, China is also the world's largest importer of tropical woods and its largest producer of cement, feeding a breakneck pace of construction. Every year, 8.5 million farmers leave their villages for fast-growing cities. The McKinsey Global Institute forecasts that by 2030, China will have 1 billion urban residents.

Although the benefits of China's economic expansion have been immense—lifting a staggering 630 million people out of poverty—so, too, has been the environmental impact. In 2006, *Forbes* magazine found that all 10 of the world's most polluted cities were in China. The water in about half of China's major waterways is unfit for drinking or even agriculture.

Even as China has invested heavily in alternative energy systems, its primary source of fuel is still overwhelmingly coal. The World Bank estimates that China's polluted water and air result in about 750,000 premature deaths each year.

Because of the country's size and influence, China's environmental concerns are no longer simply its own. China has overtaken the United States as the world's top emitter of greenhouse gases that lead to global warming. In 2006, China emitted approximately 6 billion metric tons of carbon dioxide, approximately one-fifth of the world's total.

And while China's per capita emissions are currently very low (about one-fifth that of the US), they are expected to rise significantly as an estimated 350 million people move from the countryside to the cities over the next 20 years. The Chinese Academy of Sciences predicts that during that time, the country's $CO_2$ emissions may double or more unless dramatic measures are taken.

## Who Should Pay?

Faced with such prospects—as well as a glaring international spotlight before the Copenhagen climate conference in December—the Chinese government has in recent months devoted both more rhetoric and real attention to the challenges of combating climate change.

While Beijing hopes for a global reputation boost by highlighting some green measures already undertaken, the scope of its international commitments will be motivated chiefly by domestic concerns—by the tug-of-war between political priorities and interests, a struggle often overlooked by those who see China only as an efficient top-down monolith.

As in the United States and elsewhere, business interests vie for influence, but the dynamics of the conversation are particular to China, where industry is composed of a unique mix of state-owned enterprises, foreign joint-ventures, and private companies, all with different priorities and levers for reaching government officials.

Unlike in the United States and Western Europe, where arguments between activists and climate skeptics have long defined climate discussions, there has never been much public dispute about the merits of global warming science in China.

The official 2008 government white paper, *China's Policies and Actions for Addressing Climate Change*, states without reservation: "The rise of global average temperatures since the mid-20th century is mainly caused by the increasing atmospheric concentrations of greenhouse gases, chiefly consisting of carbon dioxide, methane, and nitrous oxide, emitted as a result of human activities, such as the burning of fossil fuels and changes of land use."

Rather, the debate over climate change in China has been focused on who should take responsibility for it and who should pay for the measures to remedy it. Looming over all is the question of to what extent measures to curb $CO_2$ emissions and the use of coal could impede the country's economic growth—impediments Beijing has indicated it will not accept.

The Chinese government insists that Western countries, which have contributed the bulk of cumulative carbon emissions over time, bear the primary responsibility.

As Xie Zhenhua, vice minister of China's top economic ministry told an audience of top legislators in Beijing in August: "Developed and developing countries are still the two major factions, and the focus of disagreement remains on each country's proportion of responsibility for emission reduction, funding, and technology transfer."

## Ambitious Green Targets

Beijing has so far resisted the notion of internationally binding carbon caps, such as those that may be discussed at Copenhagen. But China does have in place two ambitious green targets, as part of its current Five-Year Plan, which would curb (though not forestall) future growth in carbon emissions.

The first goal is for China to derive at least 15 percent of all its energy from renewable sources by 2020. (The government since has talked of a more informal target of 20 percent.) The second is to reduce energy intensity per unit of GDP by 20 percent over a five-year period.

Experts have been impressed with China's green ambitions. Julian L. Wong, a senior policy analyst at the Center for American Progress in Washington, D.C., notes that China's installed wind power has doubled in each of the past four years.

John Doerr, a prominent American venture capitalist, and Jeff Immelt, the CEO of General Electric, enthused in a recent Washington Post column: "China's commitment to developing clean energy technologies and markets is breathtaking."

New York Times columnist Thomas Friedman has compared China's clean energy investments to a "new Sputnik."

Yet even as it pursues alternative energy, China will likely continue to be one of the world's leading polluters. Carbon-intensive coal, which is abundant and easily mined with cheap labor in China, is expected to supply about 70 percent of the country's energy over the next 10 years.

China's energy demand is projected to rise so steeply in coming decades that Beijing is expected to continue to build wind-farms, hydropower stations, and nuclear facilities alongside new coal-fired power plants, all in staggering numbers.

Climate change has only relatively recently emerged as a focus of government and public attention in China. Within China, local environmental problems—such as toxic factory accidents and rising cases of cancers along polluted rivers—frequently make newspaper headlines. Until the 2008 storms, though, climate change seemed a more distant and abstract concern.

As Wen Bo, a prominent environmentalist in Beijing explains: "In China, you must remember there are so many very immediate problems—environmental health, air pollution, and water quality."

## Informal Consulting, Not Lobbying

The government issued its first white paper on the potential impacts of climate change in 2007, concluding that China's vulnerability to rising sea levels and desertification was among the most severe worldwide. That finding galvanized further attention and also signaled the boundaries of acceptable public discourse—always a concern in a country with tight controls on public expression.

*Unlike in the United States and Western Europe...there has never been much public dispute about the merits of global warming science in China.*

The following year there began to be more frequent mentions of climate change in Chinese newspapers and at academic conferences, accelerating after the 2008 storms.

China today remains a nation ruled by an authoritarian one-party government. Thus, as with all matters deemed essential to the national interest, it is the country's top leadership that drives the national debate on climate change.

Going into the Copenhagen talks—which some say present China with an opportunity to improve its image in the international arena—China's negotiating position will be determined by a special inter-departmental climate committee chaired by Premier Wen Jiabao. Several ministries will be represented, most significantly the National Development and Reform Commission, China's key economic ministry.

While there is no formal and open process for lobbying the government in China—and no equivalent of Washington's "K Street"—there are opportunities for businesses and other interested parties to make their voices heard.

Most ministries consult informally with industry heads and, in some cases, with trusted academics and, very occasionally, heads of non-governmental organizations. That process has applied to the writing of recent regulations relevant to China's carbon footprint, including vehicle efficiency standards, a recent fuel tax hike, and discussions now underway in Beijing about including "carbon intensity" targets in the next Five-Year Plan.

With many of its environmental proposals, the central government has faced industry reluctance. Take the national alternative energy targets. Beijing has assigned each of China's top 10 power companies—state-owned enterprises that together account for 60 percent of China's electricity and are a significant source of carbon emissions—a goal of generating three percent of electricity from renewable sources by 2010.

But, according to calculations by Greenpeace China, based on data supplied by the companies themselves, only one appears to be on track to meet those targets.

At the root of the problem is simple economics. Coal remains the cheapest source of power generation, which hampers the implementation of China's renewable energy law.

Says Yang Ailun, climate and energy campaign manager for Greenpeace China: "The [power] companies earn more money with coal. Our argument is that the price of coal in China is too cheap. It's plentiful and easy to mine, but the current coal price doesn't reflect the serious environmental or health side-effects." (Greenpeace estimates that these "external costs" of coal burning amount to seven percent of China's GDP.)

## The Price of Power

Barbara Finamore, founder and director of the Natural Resources Defense Council's China Program, offers a similar perspective: "Right now renewable energy is a lot more expensive in China. The renewable energy law requires that utilities obtain a certain percentage from alternative energy sources, but they must do so at above market rate. The fact of matter is that it [the law] is not being well-implemented."

Ms. Finamore suggests that raising coal prices could help "level the playing field." Some groups, including Greenpeace China, call for a tax on carbon to account for the external costs that are not reflected in the current market price. But while the proposal has been around for decades, it has consistently stalled.

Among the prominent opponents of a coal tax or other price hike are the CEOs of the top 10 power companies. "Power system reform is an ongoing process. There's no need for anyone to get overexcited," Lu Qizhou, president of China Power Investment Corporation told the Chinese newspaper *Southern Weekly*. "Of course the government will make sure that the pace of reform is in line with economic development and the market's ability to cope."

Because China Power is state-owned, Lu speaks as both an industry head and a de facto part of government. As with the CEOs of all major state-owned enterprises, he was appointed by the leadership in Beijing. Previously, he served as deputy CEO of China's State Grid Corporation and then as China Power's Communist Party secretary.

In China, both the power sector and the oil sector—each critical to climate and energy policy—are dominated by a handful of large state-owned enterprises. And, in one sense, that obviates the need for lobbyists.

"There is a permanent revolving door between the Party and government and the SOEs [state-owned enterprises]," says Beijing-based political commentator Zhao Jing, who writes in the English-language press under the name of Michael Anti. "There don't need to be 'lobbyists,' when discussions can happen directly through the Party."

This interwoven network does not mean everyone always shares the same immediate goal. For instance, the power company CEOs would like to see the central government increase the price of electricity, which is currently fixed, while Beijing worries that raising prices that impact poor farmers could lead to social unrest.

However, the cozy political arrangement does mean that power companies are involved in reviewing draft regulations that would impact their sector, and they can exercise something close to veto power.

"The bosses of the big five power companies, they have official ranking—almost minister level," says Greenpeace China's Yang. "As an official of the system, they are not even made accountable to the energy bureau in the National Development and Reform Commission, which is ranked lower."

## Fueling the Future

In another sector critical to climate outcomes—transportation—the industry profile is quite different. Now the world's top auto market, China has an estimated 65 million vehicles on the road, and experts predict it will have 300 million by 2030.

But unlike the power sector, the transportation industry is not dominated by state-owned enterprises. Rather, it is characterized by joint ventures between Chinese firms and prominent US, European, Japanese, and Korean automakers, including General Motors and BMW.

Representatives from these companies are often asked for their input on draft regulations. For example, a transportation research group which develops energy and environmental policies for the Chinese government in 2008 invited comment from several foreign and domestic firms as it considered future fuel efficiency recommendations.

According to a state employee present at one of those meetings, one American company used the opportunity to oppose stricter fuel economy standards. (China's current fuel economy standards are not as stringent as those in the European Union or Japan, but are tougher than those in Australia, Canada, and the United States.) On the other hand, the employee said he believed that foreign companies had played a progressive role in helping China adopt relatively stringent tailpipe emissions standards. Most international automakers are already producing vehicles with very low tailpipe emissions for markets like California, which has among the strictest emission standards in the world. Because the technologies are already developed, these automakers can introduce them into the Chinese market relatively quickly and easily through joint ventures with domestic Chinese companies.

"If you would like to be part of it [input on regulations and green energy], you can be," says Norbert Reithofer, chairman and CEO of BMW Group, which operates a joint venture with the Chinese automotive firm Brilliance. "China is very open. But it is up to you ... If you have the right technology, you will find a partner."

As for China's historic system of state-owned enterprises, it is in many ways useful for mobilizing industry around national goals. But it also presents certain contradictions, because it enshrines established interests.

For China, envisioning the future and building from scratch—from new wind farms to megacities to future green industries—is always easier than retrofitting its past. Yet to solve its looming environmental challenges, which are now no longer its own, will require China's leadership to find a way to do both.

# The Shrinking Glaciers of Kilimanjaro: Can Global Warming Be Blamed?

By Phillip W. Mote and Georg Kaser
*American Scientist,* July 2007

The shrinking glacier is an iconic image of global climate change. Rising temperatures may reshape vegetation, but such changes are visually subtle on the landscape; by contrast, a vast glacier retreated to a fraction of its former grandeur presents stunning evidence of how climate shapes the face of the planet. Viewers of the film *An Inconvenient Truth* are startled by paired before-and-after photos of vanishing glaciers around the world. If those were not enough, the scars left behind by the retreat of these mountain-grinding giants testify to their impotence in the face of something as insubstantial as warmer air.

But the commonly heard—and generally correct—statement that glaciers are disappearing because of warming glosses over the physical processes responsible for their disappearance. Indeed, warming fails spectacularly to explain the behavior of the glaciers and plateau ice on Africa's Kilimanjaro massif, just 3 degrees south of the equator, and to a lesser extent other tropical glaciers. The disappearing ice cap of the "shining mountain," which gets a starring role in the movie, is not an appropriate poster child for global climate change. Rather, extensive field work on tropical glaciers over the past 20 years by one of us (Kaser) reveals a more nuanced and interesting story. Kilimanjaro, a trio of volcanic cones that penetrate high into the cold upper troposphere, has gained and lost ice through processes that bear only indirect connections, if any, to recent trends in global climate.

## Glacial Change

The fact that glaciers exist in the tropics at all takes some explaining. Atmospheric temperatures drop about 6.5 degrees Celsius per kilometer of altitude, so the air atop a 5,000-meter mountain can be 32.5 degrees colder than the air at sea level; thus, even in the tropics, high-mountain temperatures are generally below freezing. The climber ascending such a mountain passes first through lush tropical vegetation that gradually gives way to low shrubs, then grasses and finally a zone that is nearly devoid of vegetation because water is not available in liquid form. Tropical mountaintop temperatures vary only a little from season to season, since the sun is high in the sky at midday throughout the year. With temperatures this low, snow accumulates in ice layers and glaciers on Kilimanjaro, Mount Kenya and the Rwenzori range in East Africa, on Irian Jaya in Indonesia and especially in the Andean cordillera in South America, where 99.7 percent of the ice in tropical glaciers is found.

A simple, physically accurate way to understand the processes creating and controlling these and other glaciers is to think in terms of their energy balance and mass balance.

Mass balance is merely the difference between accumulation (mass added) and ablation (mass subtracted); in this case mass refers to water in its solid, liquid or vapor form. A glacier's mass is closely related to its volume, which can be calculated by multiplying its area by its average depth. When a glacier's volume changes, a change in length is usually the most obvious and well-documented evidence. Alaska's vanishing Muir Glacier, an extreme case, shrank more than 2 kilometers in length over the past half-century.

Glaciers never quite achieve "balance" but rather wobble like a novice tightrope walker. Sometimes a change in climate throws the glacier substantially out of balance, and its mass can take decades to reach a new equilibrium.

Added mass comes largely from the atmosphere, generally as snowfall but also as rainfall that freezes; in rare cases mass is added by riming, in which wind carries water droplets that are so cold that they freeze on contact.

The most obvious subtractive process is the runoff of melted water from a glacier surface. Another process that reduces glacial mass is sublimation, that is, the conversion of ice directly to water vapor, which can take place at temperatures well below the melting point but which requires about eight times as much energy as melting. Sublimation occurs when the moisture in the air is less than the moisture delivered from the ice surface. It is the process responsible for "freezer burn," when improperly sealed food loses moisture.

## Air, Ice and Equilibrium

Melting, sublimation and the warming of ice require energy. Energy in the high-mountain environment comes from a variety of energy fluxes that interact in complex ways. The Sun is the primary energy source, but its direct effect is limited to daytime; other limiting factors are shading and the ability of snow to reflect visible light. Energy can nevertheless reach the glacier through sensible-heat flux—the exchange of heat between a surface and the air in contact with it, in this case heat taken directly from the air in contact with the ice—and via infrared emission from the atmosphere and land surface. Energy can also leave glacier ice in several ways: sensible-heat flux from the glacier to cold air, infrared emission from snow and ice surfaces, and the "latent heat" required for water to undergo a phase change from solid to liquid (melting) or gas (sublimation).

Mountain glaciers accumulate snow at high altitudes, slide downhill—some at speeds approaching 2 meters a day—and melt at low altitudes in summertime. Some midlatitude glaciers reach sea level in part because of copious snowfall, exceeding the liquid equivalent of 3 meters per year.

Somewhere between the top and bottom of a glacier on a mountain slope, there is an elevation above which accumulation exceeds ablation and below which ablation exceeds accumulation. This is called the equilibrium line altitude or ELA. Rising air temperatures increase the sensible-heat flux from the air to the glacier surface and the infrared radiation absorbed by the glacier, so that melting is faster and is taking place over a larger portion of the glacier.

Thus rising temperatures also raise the equilibrium-line altitude. In latitudes with pronounced seasons, this expands the portion of the glacier that melts each summer and may even, in some cases, reduce the portion of the glacier that can retain mass accumulated in the winter. Virtually all glaciers in the world have receded substantially during the past 150 years, and some small ones have disappeared. Warming appears to be the primary culprit in these changes, and indeed glacial-length records have been used as a proxy for past temperatures, agreeing well with data from tree rings and other proxies.

In many respects, however, conditions are quite different for glaciers in the tropics, where temperature varies far more from morning to afternoon than from the coldest month to the warmest month. The most pronounced seasonal pattern in the tropics is the existence of one or two wet seasons, when glacial accumulation is greater and, owing to cloud cover, solar radiation is less.

Because there is almost no seasonal fluctuation in the ELA of tropical glaciers, a much smaller portion of the glacier lies below the ELA. That is, because the processes causing depletion of the glaciers operate almost every day of the year, they are effective over a much smaller area. This smaller area also means that the terminus or bottom edge of tropical glaciers tends to respond more quickly to changes in the mass balance.

An additional important distinction among tropical glaciers divides wet and dry regimes. In wet regimes, changes in air temperature are important in mass-balance calculations, but for dry regimes like East Africa, changes in atmospheric moisture are more important. Connections between such changes and global increases in greenhouse gases are more tenuous in tropical regimes. Year-to-year variability and longer-term trends in the seasonal distribution of moisture are influenced by the surface temperatures of the tropical oceans, which, in turn, are influenced by global climate. On many tropical glaciers, both the direct impact of global warming and the indirect one—changes in atmospheric moisture concentration—are responsible for the observed mass losses. The mere fact that ice is disappearing sheds no light on which mechanism is responsible. For most glaciers, detailed observations and measurements are missing, adding to the difficulty of distinguishing between the two agents.

## The Shining Mountain

What about Kilimanjaro? Tropical glacier-climate relations are different, but among them Kilimanjaro's glacial regime is unique. Its ice consists of an ice cap (up to 40 meters thick) sitting on the relatively flat summit plateau of its tallest volcanic peak, Kibo, about 5,700 to 5,800 meters above sea level and, below this, several slope glaciers. The slope glaciers extend down to about 5,200 meters (one, in a shady gully, extends to 4,800 meters). The ice cap is too thin to be deformed, and the plateau is too flat to allow for gliding. The summit's flanks are plenty steep—with angles averaging 35 degrees—but the slope glaciers move little compared with midlatitude, temperate glaciers. The slope glaciers gain and lose mass along their inclined surfaces. The plateau ice, by contrast, has two faces that each interact quite differently with the atmosphere and therefore with climate: near-horizontal surfaces and near-vertical cliffs, the latter forming the edges of the plateau ice.

What factors may explain the decline in Kilimanjaro's ice? Global warming is an obvious suspect, as it has been clearly implicated in glacial declines elsewhere, on the basis of both detailed mass-balance studies (for the few glaciers with such studies) and correlations between glacial length and air temperature (for many other glaciers). Rising air temperatures change the surface energy balance by enhancing sensible-heat transfer from atmosphere to ice, by increasing downward infrared radiation and finally by raising the ELA and hence expanding the area over which loss can occur. The first and only paper asserting that the glacier shrinkage on Kibo was associated with rising air temperatures was published in 2000 by Lonnie G. Thompson of Ohio State University and co-authors.

Another possible culprit is a decrease in accumulation combined with an increase in sublimation, both possibly driven by a change in the frequency and quantity of cloudiness and snowfall.

This argument traces its roots to 19th-century European explorers, and has been substantially improved after field work by Kaser, Douglas K. Hardy of the Climate System Research Center at the University of Massachusetts, Amherst, Tharsis Hyera and Juliana Adosi of the Twizania Meteorological Agency and others.

In 2001 Hardy had invited Kaser to join him and some television journalists in the filming of a documentary on the ice retreat on Kibo. For about a year and a half, Hardy's instruments had been deployed on the Kibo summit, measuring weather; Kaser had been studying tropical glaciers for almost a decade and a half. The team set up tents just below one of the most impressive ice cliffs that delineates the Northern Ice Field on its southern edge. During a full five days and nights on the plateau, we observed the ice and discussed the mechanisms that drive the changes, a discussion stimulated from time to time by penetrating questions from the two journalists. Kibo's volcanic ash provided a drawing board, and a ski pole served as the pencil as a picture of the regime of the glaciers on Kibo grew clearer. Thus was formed the basic hypothesis that still drives our research and that our subsequent field measurements of mass and energy balance have largely confirmed, one in which local air temperature and its changes would play only a minor role. Here is the evidence.

## Time and Temperature

Observations of Kilimanjaro's ice from about 1880 to 2003 allow us to quantify changes in area but not in mass or volume. The early European explorers Hans Meyer and Ludwig Purtscheller were the first to reach the summit in 1889. Based on their surveys and sketches, but mainly from moraines identified with aerial photographs, Henry Osmaston reconstructed (in 1989) an 1880 ice area of 20 square kilometers. In 1912, a precise 1:50,000 map based on terrestrial photogrammetry

*Comparison of historic photographs indicates that over the past century the thinning of the plateau ice has amounted to perhaps 10 meters—a rate of loss that can be explained by snowfall insufficient to balance sublimation.*

done by Edward Oehler and Fritz Klute placed the area at 12.1 square kilometers. By 2003 that area had declined to 2.5 square kilometers, a shrinkage of almost 90 percent. Much of that decline, though, had already taken place by 1953, when the area was 6.7 square kilometers (down 66 percent from 1880). Over the same period, ice movement has been almost nil on the plateau and slight on the slopes. There are indications that the slope glaciers at least are coming into equilibrium.

This pacing of change is at odds with the pace of temperature changes globally, which have been strongly upward since the 1970s after a period of stasis. Other glaciers share this pacing, with many coming into equilibrium or even advancing around the 1970s before beginning a sharp retreat.

Temperature trends are difficult to evaluate, owing to the paucity of relevant measurements, but taken together the data presented in the 2007 report from the IPCC (Intergovernmental Panel on Climate Change) suggest little trend in local temperature during the past few decades. In the East African highlands far below Kilimanjaro's peaks, temperature records suggest a warming of 0.5–0.8 degree during 1901–2005, a nontrivial amount of warming but probably larger than the warming at Kibo's peak. For the free troposphere, a deep layer including Kibo's peak, the warming rate during the period 1979–2004 for the zone 20 degrees latitude north and south of the equator was less than 0.1 degree per decade—smaller than the surface trend for that time and not statistically different from zero. Averages over a deep layer of the atmosphere, however, may be a poor estimate of the warming at Kilimanjaro's peak, although it has been argued that the warming must be nearly the same at all longitudes in the tropics, given that rotational effects are small, imposing strong dynamical constraints.

Focusing on measurements of air temperatures at the 500-millibar air-pressure level (roughly 5,500 meters altitude) from balloons, one paper suggests a warming trend in the tropical middle troposphere from about 1960 to 1979, followed by cooling from 1979 to 1997, although this study has not been updated.

Two of the data sets used to derive the tropical averages above are "re-analysis" data sets, in which observations are fed into a global dynamical model, thereby providing dynamically consistent fields of temperature, winds and so on, even where there are no observations. At the reanalysis point closest to Kilimanjaro's peak, there seems to be no trend since the late 1950s. But like the balloon and satellite data, the reanalysis data can be unsuitable for documenting trends over time.

When pieced together, these disparate lines of evidence do not suggest that any warming at Kilimanjaro's summit has been large enough to explain the disappearance of most of its ice, either during the whole 20th century or during the best-measured period, the last 25 years.

## Stuck in the Freeze

Another important observation is that the air temperatures measured at the altitude of the glaciers and ice cap on Kilimanjaro are almost always substantially below freezing (rarely above -3 degrees). Thus the air by itself cannot warm ice to melting by sensible-heat or infrared-heat flux: On the occasions when melting takes place, it is produced by solar radiation in conditions of very light wind, which allows a warm layer of air to develop just next to the ice.

A related line of evidence concerns the shape and evolution of ice. Stunning vertical walls of astonishing height (greater than 40 meters in places) tower over the visitor to Kibo's summit. These edges cannot grow horizontally but lose mass constantly to ablation (primarily to sublimation and intermittently to melting) when they are exposed to the sun—even when the air temperature is below freezing. Once developed, the near-vertical edges will retreat until the ice is gone, since no snow can accumulate on these walls.

The careful observer notes another striking fact about these walls: They are predominantly oriented in the east-west direction. This too implicates solar radiation, whose intensity is modulated by a seasonal and daily pattern of cloudiness: The daily cycle of deep convection over central Africa means that afternoons, when the Sun is to the west, are typically cloudy. The equinox seasons when the Sun is overhead are also cloudy, whereas when the Sun is to the south or north (solstices), the summit is typically cloud-free. For the same reason, the edges of the ice are retreating more slowly on the west, southwest and northwest sides.

The role of solar radiation in shaping the ice edges is evident in other features as well. As the ice retreats horizontally, it can leave behind knife-thin vertical remnants that eventually become so thin that they fall over and disintegrate. Like other explorers who came before them, Kaser and Hardy also noted the sculpted features called penitentes in the Kibo ice cap on several occasions. Penitentes are seen also in many places in the Andes and the Himalayas, where they are sometimes much larger. These finger-like features arise when initial irregularities in a flat surface result in the collection of dust in pockets, which accelerates melting in those places by enhancing absorption of solar radiation. The cups between the penitentes are protected from ventilation even as wind brushing the peaks of the developing spires enhances sublimation, which cools the surface.

If infrared radiation and sensible heat transfer were the dominant factors, these sculpted features would not long survive. Solar radiation and sublimation are sculptors; infrared radiation and sensible heat transfer are diffuse, coming equally from all directions, and so they are smoothers. The prevalence of sculpted features on Kilimanjaro's peak provides strong evidence against the role of smoothers, which are energetically closely related to air temperature.

## Mass in the Balance

What is known about the mass balance of Kibo's ice? Detailed studies of mass and energy fluxes have shown that the mass balance on Kibo's horizontal surfaces is driven by the occurrence or lack of frequent and abundant snowfall. On Kilimanjaro,

Hardy has measured the annual layering of snow directly since 2000 using snow stakes. These measurements show that the horizontal surface of the mountain's Northern Ice Field has experienced two years of near-neutral mass balance. The largest net gain observed was in 2006 when the calendar year over East Africa ended with exceptional heavy and extended rains, associated with sea-surface temperature anomalies over the Indian Ocean, and snow blanketed much of the summit of Kilimanjaro for several months.

Obviously snowfall is the main way to increase the mass of ice, but snowfall also has a role in the energy balance, one made even more important by the prominent role of solar radiation. The loss side of the balance is very much affected by the amount and even more by the frequency of snowfall: The surface of aged or polluted snow is dark and absorbs considerably more energy from solar radiation than does a white surface of fresh snow. When there is more energy available to a glacier's surface, sublimation increases. But even in below-freezing air temperatures, the same energy can increase melting if there is no wind. Meltwater from the surface is thought to be refrozen in lower ice layers; thus such melting does not necessarily constitute a loss for the ice cap as a whole. Indeed, an observer watching a slope glacier will rarely see more than a trickle of meltwater from the toe.

Comparison of historic photographs indicates that over the past century the thinning of the plateau ice has amounted to perhaps 10 meters—a rate of loss that can be explained by snowfall insufficient to balance sublimation. The observed reduction of the ice's surface area has taken place mainly at the vertical edges, however, which is not explained by snowfall patterns.

The mass balance of the slope glaciers is somewhat different from that of the plateau ice. Retreating midlatitude glaciers typically lose most mass below the ELA and little or none above. The Kibo slope glaciers, though, show shrinkage at both top and bottom. Their history suggests that in 1900 they were already far from equilibrium, but their retreat appears to be slowing; that and their convex shape suggests that they are approaching a new smaller equilibrium between the (relatively constant) loss term and the smaller accumulation term.

## Glaciers and Global Climate

The observations described above point to a combination of factors other than warming air—chiefly a drying of the surrounding air that reduced accumulation and increased ablation—as responsible for the decline of the ice on Kilimanjaro since the first observations in the 1880s. The mass balance is dominated by sublimation, which requires much more energy per unit mass than melting; this energy is supplied by solar radiation.

These processes are fairly insensitive to temperature and hence to global warming. If air temperatures were eventually to rise above freezing, sensible-heat flux and atmospheric longwave emission would take the lead from sublimation and solar radiation. Since the summit glaciers do not experience shading, all sharp-edged features would soon disappear. But the sharp-edged features have persisted for more than a century. By the time the 19th-century explorers reached Kilimanjaro's summit, vertical walls had already developed, setting in motion the loss processes that have continued to this day.

An additional clue about the pacing of ice loss comes from the water levels in nearby Lake Victoria. Long-term records and proxy evidence of lake levels indicate a substantial decline in regional precipitation at the end of the 19th century after some considerably wetter decades. Overall, the historical records available suggest that the large ice cap described by Victorian-era explorers was more likely the product of an unusually wet period than of cooler global temperatures.

If human-induced global warming has played any role in the shrinkage of Kilimanjaro's ice, it could only have joined the game quite late, after the result was already clearly decided, acting at most as an accessory, influencing the outcome indirectly. The detection and attribution studies indicating that human influence on global climate emerged sometime after 1950 reach the same conclusion about East African temperature far below the peak.

The fact that the loss of ice on Mount Kilimanjaro cannot be used as proof of global warming does not mean that the Earth is not warming. There is ample and conclusive evidence that Earth's average temperature has increased in the past 100 years, and the decline of mid- and high-latitude glaciers is a major piece of evidence. But the special conditions on Kilimanjaro make it unlike the higher-latitude mountains, whose glaciers are shrinking because of rising atmospheric temperatures. Mass- and energy-balance considerations and the shapes of features all point in the same direction, suggesting an insignificant role for atmospheric temperature in the fluctuations of Killmanjaro's ice.

It is possible, though, that there is an indirect connection between the accumulation of greenhouse gases and Kilimanjaro's disappearing ice: There is strong evidence of an association over the past 200 years or so between Indian Ocean surface temperatures and the atmospheric circulation and precipitation patterns that either feed or starve the ice on Kilimanjaro. These patterns have been starving the ice since the late 19th century—or perhaps it would be more accurate to say simply reversing the binge of ice growth in the third quarter of the 19th century. Any contribution of rising greenhouse gases to this circulation pattern necessarily emerged only in the last few decades; hence it is responsible for at most a fraction of the recent decline in ice and a much smaller fraction of the total decline.

Is Kilimanjaro's ice cap doomed? It may be. The high vertical edges of the remaining ice make a horizontal expansion of the ice cap more difficult. Although new snowfall on the ice can accumulate over the course of months or years, new snowfall on the rocky plateau usually sublimates or melts in a matter of days (with the notable exception of the period of several months of continuous snow cover in late 2006 and into 2007), partly because thin snow above dark rock cannot long survive as the loss processes reduce the reflective snow and expose the sunlight-absorbing rock. If the cap ice were much thicker and shaped in a way that allowed ice to creep outward, gentle slopes could develop along the edges; new snow would be buffered against loss and would accumulate. But steep edges do not allow such expansion.

Imagine, though, a scenario in which the atmosphere around Kilimanjaro were to warm occasionally above 0 degrees. Sensible and infrared heating of the ice surface would gradually erode the sharp comers of the ice cap; gentler slopes would quickly develop. If, in addition, precipitation increased, snow could accumulate on the slopes and permit the ice cap to grow. Ironically, substantial global warming accompanied by an increase in precipitation might be one way to save Kilimanjaro's ice. Or substantially increased snowfall, like the 2006–07 snows, could blanket the dark ash surface so thickly that the snow would not sublimate entirely before the next wet season. Once initiated, such a change could allow the ice sheet to grow. If the Kibo ice cap is vanishing or growing, reshaping itself into something different as you read this, glaciology tells us that it's unlikely to be the first or the last time.

## Bibliography

Cullen, N.J., T. Mölg, G. Kaser, K. Hussein, K. Steffen and D. R. Hardy. 2006. Kilimanjaro glaciers: Recent areal extent from satellite data and new interpretation of observed 20th century retreat rates. *Geophysical Research Letters* 33:L16502. doi:10.1029/2006GL027084

Gaffen, D. J., B. D. Santer, J. S. Boyle, J. R. Christy, N. E. Graham and R. J. Ross. 2000. Multi-decadal changes in the vertical temperature structure of the tropical troposphere. *Science* 287:1242–1245.

Kaser, G. 1999. A review of modern fluctuations of tropical glaciers. *Global and Planetary Change* 22 (1–4):93–103

Kaser, G., D. R. Hardy, T. Mölg, R. S. Bradley and T. M. Hyera. 2004. Modern glacier retreat on Kilimanjaro as evidence of climate change: Observations and facts. *International Journal of Climatology* 24:329–339. doi:10.1002/joc.1008

Mölg, T., and D. R. Hardy. 2004. Ablation and associated energy balance of a horizontal glacier surface on Kilimanjaro. *Journal of Geophysical Research* 109:D16104.

Mölg, T., D. R. Hardy and G. Kaser. 2003. Solar-radiation-maintained glacier recession on Kilimanjaro drawn from combined ice-radiation geometry modeling. *Journal of Geophysical Research* 108(D23):4731. doi:10.1029/2003JD003546

Oerlemans, J. 2005. Extracting a climate signal from 169 glacier records. *Science* 308:675–677.

Osmaston, H. 1989. Glaciers, glaciation and equilibrium line altitudes on Kilimanjaro. In *Quaternanr and Environmental Research on East African Mountains*, ed. W. C. Mahaney. Rotterdam: Brookfield, pp. 7–30.

Thompson, L. G., E. Mosley-Thompson and K. A. Henderson. 2000. Ice-core paleoclimate records in tropical South America since the Last Glacial Maximum. *Journal of Quaternary Science* 15:377–394.

Thompson, L G., et al. 2002. Kilimanjaro ice core records: Evidence of Holocene climate change in tropical Africa. *Science* 298:589–593.

Trenberth, K. E., et al. 2007. Observations: Surface and atmospheric climate change. Chapter 3 in *Climate Change 2007: The Physical Science Basis*. Contribution of Working Group 1 to the Fourth Assessment Report of the Intergovernmental Panel on Climate Change. Cambridge, U.K., and New York: Cambridge University Press.

# 3

# The Politics of Global Warming

Jason Lee/Reuters/Landov

Executive Secretary of the United Nations Framework Convention on Climate Change (UNFCCC) Christiana Figueres presses a traditional Chinese stamp as she makes her mark on a poster of the Great Climate Wall of China by Green Peace ahead of the opening session of the UN Climate Change Conference, in Tianjin October 4, 2010.

# Climate Change Policy and Economics

Scientists agree that increasing amounts of carbon dioxide in the atmosphere resulting from human industrial activity is contributing to an elevation in temperatures worldwide. Carbon dioxide produced as a by-product of burning fossil fuels traps higher than normal amounts of radiation from the sun, exacerbating the atmosphere's natural greenhouse effect. The reality of global warming presents a significant challenge to leaders worldwide. Governments and international organizations must devise and enact policies that address the problem of global warming in both the short term and the long term. Because carbon emissions occur as a result of burning fossil fuels such as oil, gas, and coal—the drivers of the global economy—the task of curbing carbon dioxide levels conflicts directly with the responsibility of governments worldwide to maintain and promote economic activity. Achieving a meaningful decrease in carbon emissions will require a balanced consideration of politics, business, and the environment.

## Fossil Fuels and the Global Economy

Policy aimed at dealing with the issue of climate change requires input from a variety of sources including climate scientists, business leaders, economists, lobbyists, and citizenry. Global warming is a problem that affects all nations. Beyond the due process and political wrangling involved in setting national policy regarding the issue lies the challenge of establishing cooperative measures between sovereign states. Competing national and economic interests makes the establishment of international regulations a difficult and time-consuming process.

Each member of the international community has different national priorities and governmental systems. They therefore have different needs to meet with respect to business, industry, and energy. China, for example, experienced unprecedented economic growth and development during the 2000s, becoming the world's largest exporter of manufactured goods. China's economic engine is fueled in large part by energy from coal-fired power plants, which generate more carbon dioxide than any other fossil fuel–based power source. Every twelve tons of coal burned as fuel produces forty-four tons of carbon dioxide, which is ejected directly into the atmosphere. Other developing economies such as those of Brazil, Russia, and India consume significant amounts of coal annually, as do countries with highly developed economies such as the United States, Japan, Germany, and South Korea.

The United States and Canada have also relied to an extent on energy from coal-fired power plants, though not on a scale similar to that of China. These nations rely instead mainly on energy obtained from hydroelectric and nuclear-fueled power plants, as well as some gas-fired power plants. Coal-fired power plants are being phased out rapidly in Canada but more slowly in the United States. The population and topography of both countries relative to their size demand an extensive transportation industry that has historically produced national economies that are in large part based on the automobile industry. Canada's energy industry includes the Alberta Tar Sands, and the United States' includes the Bakken Formation shale oil deposits as part of very active petroleum energy production. All are contributors to climate change both directly and indirectly. Canada and the United States both have political systems based on the Westminster parliamentary system of England, though with several important differences, and government policies in both countries are subject to the influence of lobbyists.

## The Environmental Impact of Global Warming

The immediate physical effects of global warming for which governments must prepare include rising sea levels in coastal areas, salinization of adjacent arable lands and potable water sources, and changing local rainfall and runoff patterns. Global warming also results in inland flooding from enhanced precipitation, agricultural losses from both flooding and drought conditions, damage and destruction of infrastructure, and increased burdens on health care systems due to excessive heat. Other impacts from increasing temperatures include cold and disaster damage, increasing strains on essential services such as fire and rescue operations, and numerous other possible effects affecting economic activity. Rising sea levels are expected to displace large numbers of people from coastal cities and other areas as flooding renders those places uninhabitable. Sea levels are expected to rise by as much as one meter by the end of the twenty-first century, which means that large areas of seaboard property throughout the world will be inundated, and some low-lying island nations may disappear beneath the waves entirely. In the nearer term, saltwater invasions will ruin arable land and potable water sources in coastal areas. Both will displace residents and affect agricultural production. This has the potential to cause significant social strife and unrest that could undermine political and social policies. Similarly, disastrous inland flooding such as has occurred in the central prairie regions of Canada and the United States has resulted in the evacuation of tens of thousands of people from their homes. Infrastructure repairs amounting to many millions of dollars are typical results of such incidents.

## Climate Change and International Relations

In the global economy, no individual government or nation exists in isolation. The atmosphere and the process of climate change recognize no boundaries, and so each nation in the international community is under threat from the problems that climate change creates across the globe. The melting of polar sea ice affects the people of Bangladesh just as it does the people of Louisiana and in much the same ways. Inasmuch as this one reality is recognized around the globe, the various nations have convened at different times to work to develop a unified stance to curtail anthropogenic climate change. The Kyoto Accord, aimed at the limitation of greenhouse gas emissions, is a prime example of that effort. There are many other national and international policies in place that aim to make improvements upon humankind's continually negative impact on the atmosphere.

That is not to say that agreement on policies such as the Kyoto Accord is universal, uncontested, or even binding. Some nations are not party to such agreements, while others have either simply ignored the stated goals of such policies or withdrawn their support under the pressure exerted by lobby groups representing diverse business interests. Many large industries have been and remain shortsighted in their manner of operations, as they work to achieve near-term profits, rather than focusing on long-term environmental concerns, and to hinder the development of policies that have the potential to threaten their profitability. Some industry lobbying groups continue to forestall or delay legislation from being enacted that would reduce the market interests of the industries that they represent.

Conversely, some businesses see the effects of climate change themselves as a market opportunity. Businesses that can help ensure the availability of those commodities put at risk by increasing temperatures—such as potable water or agriculture—stand to reap economic rewards from changing environmental conditions. Government agencies throughout the world are working to establish action plans in order to better prepare for and minimize the impacts of climate change.

The political reality of climate change is that the nations most affected by its environmental ramifications will be those least able to defend against them and those least responsible for the problem. The fifty least developed nations in the world are deemed to be responsible for only about 1 percent of anthropogenic climate change, yet they will also be the most affected by it and will have to bear a disproportionately high cost for its effects. These poor nations are also most vulnerable to the social and political upheaval that rising oceans and increasing temperatures are expected to impart. Anthropogenic climate change represents perhaps the greatest threat to humankind's continued existence on earth.

# Joint US-China Statement on Climate Change

By US Department of State, April 13, 2013

The United States of America and the People's Republic of China recognize that the increasing dangers presented by climate change measured against the inadequacy of the global response requires a more focused and urgent initiative. The two sides have been engaged in constructive discussions through various channels over several years bilaterally and multilaterally, including the UN Framework Convention on Climate Change process and the Major Economies Forum. In addition, both sides consider that the overwhelming scientific consensus regarding climate change constitutes a compelling call to action crucial to having a global impact on climate change.

The two countries took special note of the overwhelming scientific consensus about anthropogenic climate change and its worsening impacts, including the sharp rise in global average temperatures over the past century, the alarming acidification of our oceans, the rapid loss of Arctic sea ice, and the striking incidence of extreme weather events occurring all over the world. Both sides recognize that, given the latest scientific understanding of accelerating climate change and the urgent need to intensify global efforts to reduce greenhouse gas emissions, forceful, nationally appropriate action by the United States and China—including large-scale cooperative action—is more critical than ever. Such action is crucial both to contain climate change and to set the kind of powerful example that can inspire the world.

In order to achieve this goal of elevating the climate change challenge as a higher priority, the two countries will initiate a Climate Change Working Group in anticipation of the 2013 Strategic and Economic Dialogue (S&ED). In keeping with the vision shared by the leaders of the two countries, the Working Group will begin immediately to determine and finalize ways in which they can advance cooperation on technology, research, conservation, and alternative and renewable energy. They will place this initiative on a faster track through the S&ED next slated to meet this summer. The Working Group will be led by Mr. Todd Stern, US Special Envoy for Climate Change, and Mr. Xie Zhenhua, Vice Chairman, the National Development and Reform Commission. The purpose of the Climate Change Working Group will be to make preparations for the S&ED by taking stock of existing cooperation related to climate change, and the potential to enhance such efforts through the appropriate ministerial channels; and by identifying new areas for concrete, cooperative action to foster green and low-carbon economic growth, including through the use of public-private partnerships,

---

*Both sides recognize that, given the latest scientific understanding of accelerating climate change and the urgent need to intensify global efforts to reduce greenhouse gas emissions, forceful, nationally appropriate action by the United States and China—including large-scale cooperative action— is more critical than ever.*

---

where appropriate. The Climate Change Working Group should include relevant government ministries and will present its findings to the Special Representatives of the leaders for the S&ED at their upcoming meeting.

Both sides also noted the significant and mutual benefits of intensified action and cooperation on climate change, including enhanced energy security, a cleaner environment, and more abundant natural resources. They also reaffirmed that working together both in the multilateral negotiation and to advance concrete action on climate change can serve as a pillar of the bilateral relationship, build mutual trust and respect, and pave the way for a stronger overall collaboration. Both sides noted a common interest in developing and deploying new environmental and clean energy technologies that promote economic prosperity and job creation while reducing greenhouse gas emissions.

In light of previous joint statements, existing arrangements, and ongoing work, both sides agree that it is essential to enhance the scale and impact of cooperation on climate change, commensurate with the growing urgency to deal with our shared climate challenges.

# Hearing of the Select Committee on Energy Independence and Global Warming US House of Representatives

By Mario Molina

US Senate/Committee on Energy Independence
and Global Warming, May 20, 2010

Good morning, Mr. Chairman and members of the Committee. My name is Mario Molina; I am a Professor at the University of California, San Diego, and President of the Mario Molina Center for Studies in Energy and the Environment in Mexico City.

I will attempt to summarize and briefly discuss here various questions concerning the current state of knowledge related to the climate change threat, adding as well some comments on the lessons we have learned from the stratospheric ozone depletion issue that might be relevant to the climate change problem.

## Integrity of Climate Change Science

In various media reports, as well as in the Halls of Congress, some groups have stated in recent months that the basic conclusion of climate change science is no longer valid, namely, that the climate is changing as a consequence of human activities with potentially serious consequences for society. Among others, the basis of these allegations is the discovery of some errors and supposed errors in the last report of the Intergovernmental Panel on Climate Change (IPCC), which was released in 2007.

However, several groups of scientists have recently pointed out that the scientific consensus remains unchanged and has not been affected by these allegations: it is now well established that the accumulation of greenhouse gases resulting from human activities is causing the average surface temperature of the planet to rise at a rate outside of natural variability. I fully agree with this conclusion.

There are in fact some errors in the IPCC report, but they certainly do not affect the main conclusion. I will not review the nature of these errors here: they have been discussed in detail elsewhere. I just want to note that they do not appear in the Summary for Policy Makers of Working Group I, which is where the scientific consensus referred to in the previous paragraph is described in detail. On the other hand, the science of climate change has continued to evolve: new findings since

2007 indicate that the impacts of climate change are expected to be significantly more severe than previously thought. This has been documented, among others, by my colleagues at MIT and at the Scripps Institution of Oceanography.

## Uncertainties in Climate Change Science

There appears to be a gross misunderstanding of the nature of climate change science among those that have attempted to discredit it. They convey the idea that the science in question behaves like a house of cards: if you remove just one of them, the whole structure falls apart. However, this is certainly not the way the science of complex systems has evolved. A much better analogy is a jigsaw puzzle: many pieces are missing, and some might even be in the wrong place, but there is little doubt that the overall image is clear, namely, that climate change is a serious threat that needs to be urgently addressed. It is also clear that modest amounts of warming will have both positive and negative impacts, but above about 4 or 5 degrees Fahrenheit most impacts turn negative for many ecological systems, and for most nations.

The scientific community is, of course, aware that the current understanding of the science of climate change is far from perfect and that much remains to be learned, but enough is known to estimate the probabilities that certain events will take place if society continues with "business as usual" emissions of greenhouse gases. As expressed in the IPCC report, the scientific consensus is that there is at least a 9 out of 10 chance that the observed increase in global average temperature since the industrial revolution is a consequence of the increase in atmospheric concentrations of greenhouse gases caused by human activities. The existing body of climate science, while not comprehensive and with still many questions to be answered, is robust and extensive, and is based on many hundreds of studies conducted by thousands of highly trained scientists, with transparent methodologies, publication in public journals with rigorous peer review, etc. And this is precisely the information that society and decision makers in government need in order to assess the risk associated with the continued emissions of greenhouse gases. I would like to emphasize that policy decisions about climate change have to be made by society at large, and more specifically by policymakers; scientists, engineers, economists and other climate change experts should merely provide the necessary information. In my opinion, even if there is a mere 50 percent probability that the changes in climate that have taken place in recent decades are caused by human activities, society should adopt the necessary measures to reduce greenhouse emissions; but here I am speaking as an individual, not as a scientist.

In fact, recent scientific studies have pointed out that the risk of runaway or abrupt climate change increases rapidly if the average temperature increases above about 8 to 10 degrees Fahrenheit. Certain so-called "tipping points" could then be reached, resulting in practically irreversible and potentially catastrophic changes to the Earth's climate system, with devastating impacts on ecosystems and biodiversity. These changes could induce severe flood damage to urban centers and island nations as sea level rises, as well as significantly more destructive extreme weather events such as droughts and floods, etc.

## Economic Considerations

I would also like to mention that some groups have stated that society cannot afford the cost of taking the necessary steps to reduce the harmful emissions.

There are indeed significant uncertainties about the availability and costs of energy-supply and energy-end-use technologies that might be brought to bear to achieve much lower greenhouse-gas emissions than those expected on the "business as usual" trajectory. And yet, the consensus among experts is that a reasonable target to prevent dangerous interference with the climate system is to limit the average surface temperature increase above pre-industrial levels to about 4 degrees Fahrenheit; the cost is indeed significant, but only of the order of 1 to 2 percent of global GDP, and is very likely smaller than the cost associated with the negative impacts of climate change. Furthermore, besides economic considerations there is an imperative ethical reason to address the problem effectively: our generation has the responsibility to preserve an environment that will not make it unnecessarily difficult for future generations on our planet to have an environment and natural resources suitable for the continued improvement of their economic well being.

There is yet another excuse for inaction on the climate change issue that is sometimes presented by the critics, namely, that climate change is not the only problem facing society, and hence that other more urgent problems such as poverty should be addressed first. Most of us agree, of course, that there are other problems and that society should strive vigorously to achieve, for example, the Millennium Development Goals articulated by the United Nations. But it would be an error to address these problems sequentially; in fact, if some of the changes to the climate system expected to occur as a consequence of continued emissions actually materialize it will be much harder for many sectors of society to reach the desired standard of living.

> *Certain so-called "tipping points" could then be reached, resulting in practically irreversible and potentially catastrophic changes to the Earth's climate system, with devastating impacts on ecosystems and biodiversity.*

## Lessons Learned from the Montreal Protocol and the Stratospheric Ozone Depletion Issue

The global problem caused by greenhouse gas emissions has many similarities to the stratospheric ozone problem. In both cases it is crucial to change business as usual by collaboration between nations as one global community. But the quick, effective and highly successful implementation of the Montreal Protocol to protect the ozone layer stands in stark contrast to the Kyoto Protocol, the international treaty developed in 1997 to address the climate change challenge that is currently being reassessed: society has yet to find a better way to agree on effective actions on climate change.

On the other hand, the extent of change necessary to phase out the ozone-depleting chemicals was relatively small and relatively easy to monitor. The ozone-depleting chemicals (mostly CFCs) were used mainly as refrigerants, solvents and as propellants for spray cans, and could be replaced with other compounds that industry was able to develop on a relatively short time scale. In contrast, climate change is caused mainly by activities related to the production and consumption of fossil fuel energy, which has so far been essential for the functioning of our industrialized society. Effective action therefore requires a major transformation not only in a few industries, but in a great number of activities of society.

Clearly, economic development cannot continue along the same path it has followed in the past, and something has to change quite drastically. While most developed nations agree that for equity reasons they have to enable this change by providing economic resources and technology transfer to developing nations, the main problems that are being currently experienced with international negotiations result from excessive demands from some industrialized countries for "binding commitments" by all developing nations, as well as excessive demands by some developing nations for economic contributions as a condition for change. But the Montreal Protocol stands out as an example that demonstrates that an effective international agreement can indeed be negotiated. An important precedent from the Montreal Protocol is the creation of the "Multilateral Fund," which was instrumental to effectively address the stratospheric ozone question by providing resources to developing nations to achieve a smooth transition to a CFC-free society. The stratospheric ozone and the climate change problems are truly global: in the case of stratospheric ozone the nations of the world realized that they all would benefit from an effective international treaty, and that they would all lose if no agreement was reached. Thus, I believe that negotiating an effective climate change treaty is feasible, although very challenging. Nevertheless, such a treaty would undoubtedly benefit the entire world, as was the case with the Montreal Protocol.

# UN Climate Change Conference Extends Kyoto Treaty

By Karl Ritter and Michael Casey
Associated Press, December 8, 2012

Seeking to control global warming, nearly 200 countries agreed Saturday to extend the Kyoto Protocol, a treaty that limits the greenhouse gas output of some rich countries, but will cover only about 15 percent of global emissions.

The extension was adopted by a UN climate conference after hard-fought sessions and despite objections from Russia. The package of decisions also included vague promises of financing to help poor countries cope with climate change, and an affirmation of a previous decision to adopt a new global climate pact by 2015.

Though expectations were low for the two-week conference in Doha, many developing countries rejected the deal as insufficient to put the world on track to fight the rising temperatures that are shifting weather patterns, melting glaciers and raising sea levels. Some Pacific island nations see this as a threat to their existence.

"This is not where we wanted to be at the end of the meeting, I assure you," said Nauru Foreign Minister Kieren Keke, who leads an alliance of small island states. "It certainly isn't where we need to be in order to prevent islands from going under and other unimaginable impacts."

The two-decade-old UN climate talks have so far failed in their goal of reducing the carbon dioxide and other greenhouse gas emissions that a vast majority of scientists says are warming the planet.

The 1997 Kyoto Protocol, which controls the emissions of rich countries, is considered the main achievement of the negotiations, even though the US rejected it because it didn't impose any binding commitments on China and other emerging economies.

Kyoto was due to expire this year, so failing to agree on an extension would have been a major setback for the talks. Despite objections from Russia, which opposed rules limiting its use of carbon credits, the

> "At the end of the day, ministers were left with two unpalatable choices: accept an abysmally weak deal, or see the talks collapse in acrimony and despair—with no clear path forward," Meyer said.

accord was extended through 2020 to fill the gap until a wider global treaty is expected to take effect.

However, the second phase covers only about 15 percent of global emissions after Canada, Japan, New Zealand and Russia opted out.

The decisions in Doha mean that in future years, the talks can focus on the new treaty, which is supposed to apply to both rich and poor countries. It is expected to be adopted in 2015 and take effect five years later, but the details haven't been worked out yet.

US climate envoy Todd Stern highlighted one of the main challenges going forward when he said the U.S. couldn't accept a provision in the Doha deal that said the talks should be "guided" by principles laid down in the UN's framework convention for climate change.

That could be interpreted as a reference to the firewall between rich and poor countries that has guided the talks so far, but which the US and other developed countries say must be removed going forward.

"We are now on our way to the new regime," European Climate Commissioner Connie Hedegaard said. "It definitely wasn't an easy ride, but we managed to cross the bridge."

"Hopefully from here we can increase our speed," she added. "The world needs it more than ever."

The goal of the UN talks is to keep temperatures from rising more than 3.6 degrees Fahrenheit, compared to preindustrial times. Temperatures have already risen about 1.4 degrees Fahrenheit above that level, according to the latest report by the UN's top climate body.

A recent projection by the World Bank showed temperatures are on track to rise by up to 7.2 degrees Fahrenheit by the year 2100.

"For all of the nations wrestling with the new reality of climate change—which includes the United States—this meeting failed to deliver the goods," said Alden Meyer, of the Union of Concerned Scientists.

"At the end of the day, ministers were left with two unpalatable choices: accept an abysmally weak deal, or see the talks collapse in acrimony and despair—with no clear path forward," Meyer said.

Poor countries came into the talks in Doha demanding a timetable on how rich countries would scale up climate change aid for them to $100 billion annually by 2020—a general pledge that was made three years ago.

But rich nations, including the United States, members of the European Union and Japan are still grappling with the effects of a financial crisis and were not interested in detailed talks on aid in Doha.

The agreement on financing made no reference to any mid-term financing targets, just a general pledge to "identify pathways for mobilizing the scaling up of climate finance."

Small island nations scored a victory by getting the conference to adopt a text on "loss and damage," a concept that relates to damages from climate-related disasters.

Island nations under threat from rising sea levels have been pushing for some mechanism to help them cope with such natural catastrophes.

# Scientists, Policymakers, and a Climate of Uncertainty

By Fred Powledge
*BioScience,* January 1, 2012

## Can Research Gain a Foothold in the Politics of Climate Change?

Climate change—fears of it; predictions about it; proposed reactions to it; and in some cases, denials of it—is a momentous issue for humanity, and it promises to become even more important. However, despite its importance—and mountains of scientific research on the topic—real progress by those with the means to address climate change has been slow in coming.

The warming climate, with its accompanying intense weather, rising seas, and wrenching changes in agriculture, human health, and ecosystems in general, is not a recent discovery. Scientists and policymakers have been talking about it for decades. Climate scientists have produced reams of information about the change that is upon us—some showing that the shift is well under way, some offering ideas about how to cope with it by mitigating its effects or by adapting to it. Scientists are starting to develop the tools necessary to discover links between global change and specific local events, such as floods and droughts.

Much of the US-based research is directed at people who make political decisions: members of Congress, the president, state legislators, municipal officials. But whereas the flow of information from scientists has been copious, the response from policymakers, including those at the topmost positions in government, has been minuscule. Interest in preparing for climate change seems to have been superseded by concerns about the economy. But the problems of a changed climate march on, and at some point, policymakers will have to deal with them.

As far back as 1978, Congress passed the National Climate Program Act in response to a need "to assist in the understanding and response to natural and man-induced climate processes and their implications." The legislation stated that "Congress finds and declares" that "an ability to anticipate natural and man-induced changes in climate would contribute to the soundness of policy decisions in the public and private sectors," and that "the United States lacks a well-defined and coordinated program in climate-related research, monitoring, assessment of effects, and information utilization." Today, after dozens of data-heavy reports on climate change, bolstered by the work of the Intergovernmental Panel on Climate Change

(IPCC), the crisis remains unaddressed, in large part because the issue has become a partisan arguing point—increasingly so as a presidential election looms. In April 2011, the House of Representatives considered action stating that "Congress accepts the scientific findings...that climate change is occurring, is caused largely by human activities, and poses significant risks for public health and welfare." The national legislators voted the proposition down 240 to 184, largely along party lines.

Scientists who study climate change, then, are left with the following question: How do we communicate in a meaningful way with policymakers?

First of all, says Pamela A. Matson, an interdisciplinary earth scientist at Stanford University, scientists should not tell policymakers what to do. Matson chaired one of four expert panels charged by the US National Academies in 2009 with studying climate change and recommending responses to it. The panels—Limiting the Magnitude of Future Climate Change, Adapting to the Impacts of Climate Change, Informing an Effective Response to Climate Change, and Advancing the Science of Climate Change—each produced extensive reports. Together, these reports and a summary volume form a solid tutorial on the challenges facing the nation . . . . Matson headed the Advancing the Science of Climate Change panel.

"We can't do much about politics," said Matson in an interview, "other than to keep reminding people that there is a huge amount of evidence about climate change, its causes, and the risks associated with it. Decisionmakers of all sorts will need to decide if they want to take the risks and expose future generations to even greater risks, but scientists can do our best to provide the best and most clear information about those risks and uncertainties. We can also help by providing viable options for reducing those risks—options that make sense from social, economic, technical, and environmental perspectives. The Advancing report calls for more research designed to develop such options and thus support decisionmakers who want to respond to the risks of climate change by limiting it and adapting to it."

In practically all of the heavyweight studies of climate change, including those in the National Academies quartet, the panelists assumed that the federal government is the logical leader in climate policymaking, if only because climate does not respect local boundaries (or international ones either). But there are scant signs of leadership from federal officials, and there is active opposition to any climate action (or even the notion that climate change exists) among some national legislators.

Progress at the federal level, says Peter Raven, cochair of the National Academies panel Informing Decisions and Actions, is missing. His panel recommended that the federal government create "comprehensive, robust, and credible information systems to inform climate choices and evaluate their effectiveness." "My impression," said Raven recently, "is that it hasn't happened. The United States is the only nation in the world where serious doubt is expressed about the scientific conclusion that the climate is warming rapidly and that human beings are the principal underlying factor. That kind of anti-intellectualism can only hurt as we try to maintain our standing in science and technology against some very stiff competition internationally."

Robert Fri, visiting scholar at Resources for the Future, chaired the Limiting the Magnitude of Future Climate Change panel, which recommended "a framework of national goals and policies" to limit greenhouse gases. Asked about the report's reception, he replied, "Policy attention has been focused on the economy recently. Not much has happened on the climate front as a result."

So the search for leadership has turned elsewhere—to local, state, and regional governments and private organizations. Several states have undertaken serious studies of climate change problems, many of them as a result of gubernatorial executive orders issued a few years ago, when climate change was considered less controversial. Some went a step further and set up commissions to recommend legislative action. The results of those efforts have been mixed; the economic crisis has sapped much of the climate change energy of local, as well as national, politicians.

Arizona's Policy on Climate Change, instituted on 2 February 2010, under Governor Jan Brewer, seeks to reduce greenhouse gas emissions "while maintaining Arizona's economic growth and competitiveness." Yet in the same executive order that established the policy, Brewer pulled her state out of a promising regional effort, the Western Climate Initiative, to limit the gases. The 15-member commission created by Brewer's executive order included representatives of electric power, manufacturing, and mining companies, but no scientists.

The Colorado Climate Project created a blue-ribbon action panel in 2007 that, a year later, produced 70 recommendations for reducing greenhouse gas emissions and preparing for the coming changes. The panel's efforts appear to have been received favorably by policymakers.

Texas, beset by wildfires in 2011 that some experts linked to climate change, has no state adaptation plan, according to a survey by the Pew Center on Global Climate Change. Texas leads the states in carbon dioxide emissions.

In Connecticut, the Governor's Steering Committee on Climate Change says that it is "working with stakeholders" and "assessing the impacts of climate change." However, the committee's Web site devotes approximately half of its space to thanking the company that designed its logo.

Massachusetts has made a more strenuous effort. In September 2011, the state sent a report to its legislature in which the expected impacts were analyzed and suggestions were made for reactions to them. The proof of the pudding in Massachusetts, as in the nation as a whole, lies in what happens after the recommendations go to the policymakers. In the Bay State, that would include the office of Frank I. Smizik, the chair of the House Committee on Climate Change.

Smizik believes that climate change is going to affect "every corner of our lives," is already creating problems, and is "one of the most complex and interrelated problems we face. And that makes it even more of a challenge to confront." In an interview, the legislator praised the state's framework report both for its "long-term and far-reaching strategies" and for the "small, incremental improvements we can make at little or no cost. This aspect of the report makes the necessary task of climate adaptation seem more feasible."

Does this mean that Massachusetts's policymakers will rush to enact laws to deal with the well-documented menace? "Ideally, the need to adapt to climate change will be taken seriously, and these strategies will be swiftly transformed into reality in Massachusetts," said Smizik. "Realistically, it won't be that easy. My job as the Chair of the House Committee on Global Warming and Climate Change is to advocate for the strategies that I think are not only most effective but also the most politically feasible at this time. I could see some of the more short-term, cost-effective strategies making progress in the State House as legislative packages. Meanwhile, some strategies will be directly implemented by state agencies.

"I feel these issues are very pressing and they warrant our immediate attention. But… the legislating and governing process takes time, with good reason. The release of this adaptation report is an important step for our state, but it's hard to say how quickly or slowly these adaptation measures will become law. The economic climate is such that legislators are very focused on job creation and the like, and environmental issues are often pushed to the back burner in this situation."

The state of Washington is another place where policymakers take climate change seriously. Hedia Adelsman is the director of the state's Department of Ecology. Her predecessor in that job was Christine Gregoire, who went on to become the state's attorney general and then governor. Adelsman credits Gregoire with helping move climate awareness into action. Also, the University of Washington produced an exhaustive examination of climate change and ways to react to it that has served as a vital resource for legislators and other policymakers.

Even though Washington is more environmentally conscious than most other states, economic troubles have hampered action on the climate front. "The conversation is not as active as it used to be," said Adelsman in an interview, "[for] a couple of reasons: One of them—no surprise—is all the budget problems that the state is facing. And when you talk about the impact of climate change, people really see it as in the future. And right now we have all these urgent problems facing us. Some of the impacts that would be due to climate change… people don't see as something that's happening now. They see it as happening a little bit more in the future. So they feel like they can get back to it later… But the communication continues.

"It all depends on how you describe it," she said. "If you are describing the problem as a water-availability issue—if we are going to face a major issue with water available for irrigation; for municipal, for our fish, the salmon—you get people to listen. If you talk about it as global warming, it's immediately, 'go away.'

"It's kind of sad to not call it by what it is, but it's much easier to communicate with the public and with the policymaker if you put it in these terms."

Some, including the authors of the National Research Council's panel Advancing the Science of Climate Change, think that a likely source of federal leadership can come, with a few modifications, from an existing organization: the US Global Change Research Program (USGCRP), which was created in 1989 as a presidential initiative by George H. W. Bush. USGCRP coordinates and integrates the global change work of 13 agencies, from the US Department of Agriculture to the US Department of Defense to the US Environmental Protection Agency and the Smithsonian Institution.

Thomas Armstrong, of the White House Office of Science and Technology Policy, is USGCRP's executive director. In an interview, he said that he certainly agreed with the panels' findings. "But the real challenge comes not when you recommend strategic changes but when you get down to the nuts and bolts of implementing programs." One potential obstacle, he said, could come if 13 agencies, which together receive $2.8 billion in global change research funds, are asked not only to funnel their work through a single office—his—but are made to redirect their agencies' funding as well.

"If we were to take this level of enterprise and now say that it is no longer just coordinated but actually run out of my office, with all the resources coming to my office, that would be a big change in our collective way of doing business, to say the least," said Armstrong. "USGCRP is the 13 agencies, and the 13 agencies are USGCRP. The current confederation of 13 agencies gives them ownership and a true piece of leadership of this remarkable program. I would be concerned that in redirecting their fiscal resources, we would be taking away the ownership and investment of 13 federal agencies who use their research and operational funds… to meet 13 agencies' mission mandates. In effect, we would be taking the USGCRP out of what makes this program so strong: the agencies themselves."

## The Middle Communicators

Science faces several hurdles in delivering climate information to policymakers. An obvious one is the climate change deniers, who can produce distrust of all science in the minds of some lawmakers. Another is the difficulty of conveying levels of scientific uncertainty, which scientists handle by expressing their projections, as does the IPCC, in degrees of confidence, such as very high confidence (a 9 out of 10 chance), medium confidence (5 out of 10), very low confidence (less than 1 out of 10), and so on.

To deal with communication matters such as these and many more, there has arisen a system of "middlemen," or for a better term, "middle communicators." These can be government agencies, such as the groups that produced the state action plans, or state agencies such as Hedia Adelsman's in Washington, and federal executive-branch agencies such as the USGCRF itself. Or they can be quasiacademic, quasigovernmental organizations, such as the eight regional Climate Science Centers that were established in 2009 by the US Department of the Interior. The centers are led by the US Geological Survey and hosted by one or more universities.

Another significant area of communication is provided by private foundations and nongovernmental organizations—among them, the Pew Center on Global Climate Change, which has been working since 1998 to gather and distribute information on

> *Smizik believes that climate change is going to affect "every corner of our lives," is already creating problems, and is "one of the most complex and interrelated problems we face. . . ."*

climate change. Another, CAKE (for Climate Adaptation Knowledge Exchange), an undertaking of Island Press and EcoAdapt, maintains a "shared knowledge base for managing natural systems in the face of rapid climate change." The database, which contains a large compendium of tips for communities seeking techniques for adaptation, contains dozens of case studies from around North America.

The middle communicators appear to be having some success in getting scientific information into the hands of those who already want to use it in their decisionmaking. Daniel Ferguson, program director at the Tucson-based Climate Assessment for the Southwest, part of a regional sciences program run by the National Oceanic and Atmospheric Administration, says that it helps that "we are more commonly working with decisionmakers much closer to the ground in terms of decisions that may be able to utilize scientific insights for something more akin to real-time decisionmaking. . . . We certainly find people who are actively seeking out scientific information to make decisions that help them deal with climate." Many of these decisions relate to the drought that has enveloped the region for years and are vital to policymakers who deal with water availability.

Ferguson's organization does a lot of its communicating when public and private groups, including tribal organizations, invite a staff member to give a talk. Workshops are also useful for spreading scientific knowledge about adaptation and mitigation. To a great extent, then, Ferguson's audience is presold on the idea that climate change is important. These are "interested parties who have, for one reason or another, recognized climate variability or climate change as something they really need to learn more about or . . . factor into their operations. Sometimes it's because there's a mandate to pay attention to climate"; since 2009, agencies of the US Department of the Interior have been under secretarial orders now to make climate change part of their work. "Sometimes it's because an individual feels like their organization is especially vulnerable to the impacts of climate, and they want to learn more so they can begin to address this internally. Sometimes it's because an organization or government agency has observed a disturbing climate impact that forces them into action." The current drought is an example of this, as is forest mortality in the western United States and Canada.

But on many other levels, the science of climate change has had a much rougher reception. North Carolina was one of several states that became active in the issue of climate change around 2005. State lawmakers passed, and then-Governor Mike Easley signed, a bill establishing the 34-member Legislative Commission on Global Climate Change and charging it with addressing the problems of global warming, as well as with looking into ways to mitigate the threats and with investigating the potential of carbon markets. The commission worked for five years and came up with 42 recommendations. At the end of that time, the North Carolina legislature failed to allow the commission to continue its work.

State Representative Pricey Harrison, who was cochair of the commission, now says that industry clout doomed the effort. "Industry pressures got in the way of the scientific evidence and stifled any momentum for doing anything about North Carolina's pursuing a low-carbon future," she said in an e-mail interview. "We had a difficult time from the start with the automobile industry, investor-owned utilities, forestry association, Farm Bureau, manufacturers, Chamber of Commerce, [and so on]. They all teamed up and fought every effort to recommend anything of substance. We had a very large stakeholder group convened, and it was quite balanced, so it was difficult to operate without consensus." Even with consensus, she said, anything that did make it out of her committee "was DOA at the legislature."

Harrison says that strong opposition to progress came from conservative and libertarian groups such as the John Locke Foundation and conservative philanthropist Art Pope, who Harrison called "our own version of the Koch Brothers." (The Koch brothers, wealthy US industrialists, are major funders of conservative and libertarian causes.)

"It was quite frustrating," she said. "I could barely get an innocuous bill passed that required that our state agencies review the impact of climate change and make recommendations as to policy changes and other procedures. There wasn't even any appetite to deal with adaptation—because I guess that meant we would be acknowledging that climate change is an issue in North Carolina.

"North Carolina will see little to no activity to address climate change while the current leadership is in place. Unfortunately. This is particularly troubling, since we are the third most vulnerable coast in the country to sea-level rise," she said. In addition, the state is "ground zero" for hurricanes and has also experienced "terrible droughts" and "crazy snowfalls." Moreover, Harrison said, "We are one of the biggest emitters of greenhouse gas[es], at around 24th in the world—more than many similarly sized countries."

She added that the state has managed to pass a number of clean-energy policies and has effectively reduced its carbon footprint. The selling point for those measures was jobs. For climate change legislation to advance on its own merits, she said, "I am afraid we will need federal leadership."

## National Research Council Definitions

Climate change, in the United States and elsewhere, covers most areas of the human experience. A report from the US National Research Council lists the following areas of concern: changes in the climate system; sea-level rise and its effects on the coastal environment; freshwater resources; ecosystems, ecosystem services, and biodiversity; agriculture and aquaculture; public health; cities and the "built environment"; transportation systems; energy systems; solar radiation management; national and human security; and climate policy.

# A Life Measured in Heartbeats

By Brian Cox and Robin Ince
*New Statesman,* December 21, 2012

*For 60 years, David Attenborough has been a champion of science. Does he feel it is now under threat?*

**Robin Ince:** No other individual is held in such awe by as broad a group of people as Sir David Attenborough. On seeing him, one eloquent friend felt he must say something, and so he bounded up, blurted out "Thank you," then scarpered. I have seen people held in high regard reduced to gibbering fan-kids on finding themselves in the same room as him. After 60 years in broadcasting, a career that has included commissioning Kenneth Clark's *Civilisation* and Jacob Bronowski's *The Ascent of Man*, as well as astounding investigations into the varieties of life on this planet, he continues to work every day of the year with the occasional exception of Christmas Day.

The landmark series *Life on Earth* was my introduction to the theory of evolution and the work of Charles Darwin, a man who increasingly fascinates me the older I become. The writings of Darwin convey a mind restlessly attempting to understand the life he sees before him and driven to explain why it seems to be as it is. David Attenborough has allowed us to stay in our armchair and dwell on the complexity of living things on this small but densely populated planet. Contemplating the seeming rarity of life in the known universe, he once described the earth as "a meadow in the sky."

Whenever the case against television is brought up, the work of Attenborough is called by the defense. In the television world, where so much is required to be fake, from the smiles to the feigned interest of the interviewer, Attenborough conveys passion, a wish to communicate not defined by pay packet or celebrity. He is not making a film about tribal art or bowerbirds or environmental crisis because it's a job; he is doing it to share ideas, convey wonder and to learn for himself. This is not a tired academic going through the rigmarole of explaining life one more time; this is someone able to capture the excitement of the adventure because he is still on it. Where cynicism and ironic distance can seem the way of the 21st century, here is an unashamed enthusiast. As he leaned forward during this interview and told us of seeing the hasty and flamboyant mating ritual of a hummingbird slowed down so each intricate detail could be examined, he reminded me that it is criminal to feel bored in a world so rich.

Of walking in the Brazilian jungle, Darwin wrote: "The delight one experiences in such times bewilders the mind. If the eye attempts to follow the flight of a gaudy butterfly, it is arrested by some strange tree or fruit; if watching an insect, one forgets it in the stranger flower it is crawling over The mind is a chaos of delight." David Attenborough has helped stimulate minds into that state of chaotic delight.

**Brian Cox** How has the public perception and presentation of science changed over the 60 years that you've been involved with it?

**David Attenborough** I suppose there was a mystique about science and also a certain suspicion of it. I went to grammar school: if you were very bright, you did classics; if you were pretty thick, you did woodwork; and if you were neither of those poles, you did science. The number of kids in my school who did science because they were excited by the notion of science was pretty small. You were allocated to those things, you weren't asked.

This was in the late 1930s/early 1940s, and schools are better about that now. But it's indicative of the way people thought about science: they didn't really connect science and technology—"OK, yes, that's how you mend a fuse, but that's not science"—which of course is nonsense.

Science was seen as something more remote and less to do with everyday life. Since then, our society has become so technologically based that you really can't be a fully operating citizen unless you understand basic science. How are you supposed to make judgments about the health of your children if you don't believe in science? How are you supposed to make a judgment about a generation of fuel and power if you don't believe in science? You can't operate as a sensible voting member of a democratic society these days unless you understand fundamental scientific principles to a degree.

**BC** And yet, today, there seems to be a politicization and a polarization of ideas. There's a certain camp that distrusts science, particularly when we start talking about vaccination policy or climate change.

**DA** I've never known a time when scientists weren't hypersensitive about not being understood by the rest of society. Whenever I come across scientific institutions, unless they are absolutely remote, sort of dealing with little pockets of independence—as hobbies, really—they never think that society as a whole understands what they are about. But I think it's not as bad as many scientists now believe; I think it was worse 50 years ago.

**BC** Do you think the presentation of science on television has a role to play?

**DA** Yes, and the BBC can be reasonably proud there. In 1952, when I joined, there was a chap whose responsibility was "science" and he was a physicist. The head of the department where I worked—rather absurdly called Talks—was an Oxford geneticist, and she was extremely keen on science.

One of the great achievements of the 1950s and 1960s was a series called *Your Life in Their Hands*, which dealt with medical science. It presented the scientific evidence for the connection between tobacco and cancer, against the entrenched opposition, all of which you can quite easily imagine.

In those days, the doctors concerned were not allowed to use their names—it was against medical practice and ethics—so he was just called the "television doctor" and he presented the evidence of the connection between the two over and over again on television. A lot of people tried to stop it, but he carried on. It ruined his career, I suspect, in the medical sense, but he stuck to his guns. It's one of early television's badges of honor.

**BC** Do you feel you have become more polemical recently? Because I know you have written that the real delight you find is essentially in observing nature.

Have you become more a campaigner?

**DA** It's OK for you—you are a scientist. I am not a scientist.

**BC** You are a fellow of the Royal Society.

**DA** But I'm an FRS for popularizing science and I don't have a right to go out and talk about the details and physics of the upper atmosphere. I only take it from what other people say. I respect researchers who do that sort of work and can make some degree of assessment as to whether they're cranks or not.

If you appear as frequently on television as I do, looking at bunny rabbits or whatever, it's very tempting to think people notice what you say, because you're the one person they see and hear on the subject of science, so they think you're a scientist. You are not a scientist. I'm a television journalist and mustn't become so intoxicated with my own intellect that I suddenly believe that I discovered these things or I have a privileged position to assess them.

All I can do is use what scientific education I have to say: "Look, listen, I know this about this, just look at this graph."

In the early days of dealing with climate change, I wouldn't go out on a limb one way or another, because I don't have the qualifications there. But I do have the qualifications to measure the scientific community and see what the consensus is about climate change.

I remember the moment when I suddenly thought it was incontrovertible. There was a lecture given by a distinguished American expert in atmospheric science and he showed a series of graphs about the temperature changes in the upper atmosphere. He plotted time against population growth and industrialization. It was incontrovertible, and once you think it's really totally incontrovertible, then you have a responsibility to say so.

**RI** With *Frozen Planet*, there was some hullabaloo over the editing-together of footage of a polar bear birth. And journalists turned that into: "Can you trust anything in this series?" Do you ignore that kind of thing?

**DA** No, you can't ignore it and you shouldn't ignore it—because people are reading it, so you should respond.

With that, I wouldn't say it's mischievous, but you know, the fact that it was shot in the zoo was made clear by us to start with. If the program had shown me, or anybody, saying, "Here I am, trudging across the Arctic, and I am going to see if I can find a polar bear den and then crawl through snow and then put it in a shot"—well, that's a lie.

But if you are trying to create a sequence in which you are making a genuine attempt to explain the biology of bears, one of the key things is that the young are born in the middle of the hibernation period. Unless you understand that, it doesn't make sense of the life cycle. You are doing a dud program if you don't include that sort of thing.

So we didn't do the drama of showing me crawling around. We said: "This is the polar bear's life." I think I got one letter saying that was wrong, and all the rest—several dozen letters—were from people saying, "What do they take us viewers for? Do they think we're stupid enough to think that every shot you've ever taken is exact?"

**RI** Your new *Sky* series is about Galapagos. Was your first visit to Galapagos with *Life on Earth*?

**DA** Yes.

**RI** It's seen now, in terms of biology, as almost a mystical place, because of Charles Darwin. He used to talk about his mind being a "chaos of delight" there— did you find that, too?

**DA** Yes, before I ever went, if you talked about the Galapagos and Darwin, most biologists and naturalists would think of finches—and quite right, too: that's what clinched the thing as far as Darwin was concerned.

But it wasn't the finches that put the idea [of natural selection] in Darwin's head, it was the tortoises. The reason he didn't use the tortoises [in writing *On the Origin of Species*] was that, when he got back, he found he didn't have localities on the tortoise specimens. Here the great god, the greatest naturalist we have records of, made a mistake. His fieldwork wasn't absolutely perfect.

That was quite entertaining to start with, but also it's much more comprehensible for the audience that [in *Life on Earth*] we went with the tortoise and the length of its neck rather than less dramatic changes in the shape of the bill of a sparrow-like finch.

**RI** Now that you have been making shows for 60 years, are there still moments when the natural world still flabbergasts you?

**DA** Things flabbergast me all the time. I wasn't involved in filming it, but a friend of mine was up in the Andes filming the courtship display of a particular humming-bird, a high-altitude hummingbird. The female was trying to advance, and the male was coming and going, "prrrrrt," and then it was gone.

You think, "Oh, that's OK"—but then my pal had the wit to shoot it at 250 frames [a second] and you suddenly saw the complexity of the display. It was astounding, all at this very, very high speed.

The moment you say that, you think of the timescale of hummingbirds, the speed of their hearts and the temperatures at which they operate. Their timescales are quite different. You suddenly realize your own limitations: how your sensory perceptions are governed by your heart rate.

I thought that was so exciting, and it taught you so much, not only about how complex nature is, but about how impoverished your perceptions can be, governed as they are by your human condition.

**BC** I suppose you would use "impoverished" in terms of not being exposed to these ideas and not being able to see the behavior of these animals. So how important do you think broadcasting, and specifically public-service broadcasting, is to the fabric of the country?

**DA** I think you and I know, because we've spoken together before. You and I feel the same way about this. It seems to be a fundamental responsibility of a broadcaster to make these things apparent.

As we were saying earlier, how can you be a competent member of a democratic society, with a right to vote, and have no conception of or no basis for understanding the technology on which the whole of your society is built, from the food you eat and the schools you go to, to the way you move, the way you communicate, the way you look after your kids, the way you read at night? All these decisions are fundamentally scientifically based.

Now, if a public-service broadcasting organization has any responsibility, then it must be in these areas where it doesn't necessarily make a financial profit to talk about it.

Science is an obvious example. We spend a lot of time saying, "Oh yes, it's very accessible and very exciting"—but it's not always accessible, it's not always exciting, and it's quite easy to be boring about. It is the job and responsibility of the broadcaster to deal with these problems and make sure it's available to everybody. I really feel very powerfully about that. If the BBC as a public-service organization allowed its scientific output to dwindle, then it ought to be a national scandal.

**RI** Is there a danger of underestimating the public? Is there a worry that television producers get so worried about viewing figures that they begin to think their audiences are more stupid than they are—and decide to give them more dance shows, cookery, or whatever?

**DA** It's certainly a danger, there's no question about that. Perish the thought that the BBC as a public-service organization should suppose that the only criterion for success is audience size. That would be absolutely dreadful.

In my view, the proper attitude of a public-service broadcaster is that it should attempt to cover as broad as possible a spectrum of human interest and should measure success by the width of those views. There shouldn't be all that large a number of gaps in the spectrum; and a major element in the spectrum is scientific understanding. The fact that it doesn't necessarily get as big an audience as cookery is of no consequence.

**BC** We touched on this at the start—over the time that you've been in television, making natural history programs and science programs, it's very tempting to see a degradation in their quality. You read that a lot in the newspapers: "It's not as good as

> *In the early days of dealing with climate change, I wouldn't go out on a limb one way or another, because I don't have the qualifications there. But I do have the qualifications to measure the scientific community and see what the consensus is about climate change.*

it used to be." So are you optimistic or pessimistic? Do you think standards are improving or declining?

**DA** I think that standards vary and there will be cycles. There were in my time, and it was certainly the case when ITV came into existence. I was at the BBC when it was a monopoly—and we assumed that because viewers had nothing else to look at, they would watch us and we were doing OK.

It came as a great shock to us that this wasn't the case—that when ITV came into existence, suddenly lots of viewers who had the choice moved elsewhere. That was a very salutary lesson. You have to recognize that you don't just put programs out there to the poor public, who are supposed to regard themselves as privileged to have it put in front of them. You've got to do better than that, you've got to proselytize, if you believe that the standards of civilization, of our society, are worth purveying.

**BC**  Now you are working with Sky. Did you think you'd ever move to a commercial station?

**DA** No.

**BC** Does it hearten you that a commercial station will make these programs and put this amount of money into them?

**DA** The division that happened in 1954 when ITV came into existence seemed very clear, and it was exaggerated on both sides. We were way up in the ivory tower and didn't give a damn about what the audience thought, and we thought the other side were just muckraking.

Well, the broadcasting landscape has become much more complex, much more sophisticated than that. There's no longer a duopoly; there are lots of us doing all sorts of things and I would never have thought, for example, that the BBC would take advertising.

You would say: "It doesn't, does it?" But the fact is that it does. UKTV, in which the BBC has a 50 percent share, takes BBC programs and puts advertisements in them. The first time I saw that done to one of mine was a bit of a shock. I thought, "I didn't sign up for this."

ITV is not as black as it was painted, and the fact that Sky is prepared to do a 3D program and one of the first important things based on vertebrate paleontology—that's not bad.

**RI** In your 60 years making programs, was there a moment when you thought "the childhood me could not imagine that I would be in such a situation"?

**DA** Oh, yes. Quite a lot, really. It occurs over and over again. But I remember one particular occasion in northern Australia. We built a hide on a big billabong and got there at about three o'clock in the morning, a couple of hours before sunrise. And the sun comes up, and you see this billabong thronged with magpies, geese, herons, cockatoos, kangaroos, coming down to drink, marine crocodiles.

You had a vision of the natural world, a Rousseau-esque kind of thing. You suddenly held your breath, because you were in a strange, godlike thing; you saw the world as it was without humanity in it. And then suddenly something happened— I forget what it was, someone made a noise or something—so the whole thing was gone. But that was a moment of perception which haunts you.

**BC** This edition of the *New Statesman* is a celebration of reason, and I think maybe science is on the rise again. Are you optimistic or pessimistic about our country's scientific future?

**DA** I would have thought that, among the younger generation, the dominance of technology is now such that you have to engage with realities of science. Just by working with some of these technological devices you are learning binary mathematics, you are learning logical processes which you didn't have to do when you were considering what Ovid wrote or Plato's arguments.

I begin to hope that this aspect of technology will lead to a rational approach to thinking about problems we're going to face. That will be very important to instill in the kids. Do you think the same?

**BC** Yes, to an extent. Although earlier I asked the question about polarization—because there is a polarization and a politicization, particularly on climate change, I'm thinking. There are people you may credit as being intelligent, people who we don't have to name—we've got Nigel Lawson, for example

**DA** But you're missing one element, because you're talking as if the playing field is even. It's not. The playing field is in trying to see things as they are and, on the other side, trying to see things as you wish they would be because that would give you a bigger profit. Then you test the veracity or the strength with which you believe in rational thought.

So yes, two and two is four, but if saying two and two is four and a half would be worth a couple of hundred pounds, then you say it. That said, the theory of numbers itself is very, very interesting. [Laughter] It's a generalization in many ways. If you look at it from one point of view, it could well be four and a half.

**BC** Do you think we have a chance as a society of becoming so rational, or at least so respectful of science, that we will be able to address those challenges?

**DA** I hope we will, but you know the problems. As with the tobacco industry that we were talking about before—the leaders, were they all absolutely duplicitous? Or did they say, "I've become a multimillionaire as a consequence of this, and it's not actually proven, this connection between cancer and nicotine. A lot of people love it, and the third world, what would they do without tobacco?"

You fudge it out of self-interest. I must be careful, because there are some people who genuinely believe it; quite a lot of people find it easier, happier, softer, more comfortable to believe than to face the awkward reality.

# Global Warming Conference: Bill McKibben's Remarks

### By Bill McKibben
### US Senate, March 16, 2013

It's very good to be here, and I feel like this is the answer to a kind of riddle. What would draw 500 Vermonters inside on the prettiest winter Saturday? And the answer is Bernie Sanders.

One of the things that my life has involved, that has been kind of different for me in the last five or six years, is I've had the opportunity to meet a lot of politicians. Senator Sanders is different from the other politicians for the most part.

Let me just say there has no greater champion on these issues in DC, and we're going to see it again next week when some people in the US Senate are going to try and do a kind of backdoor approval of this Keystone pipeline and I have no doubt that Bernie will be in the absolute lead of making sure it does not happen.

Look, most of the hard work's been taken care of already for me today. The young people did an amazing job of laying out where we are with climate change and what's going on. I'm really not going to bother to talk about that at all. You know, I wrote the first book about all this almost 25 years ago now, and the only thing to say is that it's happening faster and on a larger scale than we thought.

Last summer was very scary. By the time the summer was over, the Arctic had, in essence, melted. Eighty percent of the ice that was there 40 years ago is gone. We've taken one of the largest physical features on Earth, and we've broken it. And that's not a good sign, when you start breaking your major physical features. That's the pace at which things are happening.

And we can talk all day and all night about how warm it's been in the United States, and we can talk about Hurricane Irene and about Hurricane Sandy. This is what's happening now, and this is what happens when you raise the temperature 1 degree. The same scientists who told us that would happen, have told us that unless we get our act together very fast, it's going to be four or five degrees before the youngest people in this room are as old as I am.

So that's our job, to make sure that it does not happen. And it's going to be a hard fight, but the real news I want to bring you guys today, the little part I want to do, is to basically just to introduce you to who your brothers and sisters around the world are, who are taking part in this fight, just so you have a sense, that though Vermont is often in the lead, Vermont is not alone in these fights.

Six years ago now, and there's people in this room who were a part of it, we walked up the west side of Vermont, up Route 7, from Robert Frost's old summer writing cabin up to Burlington. We took five days, and we slept in fields, and since I'm a Methodist Sunday school teacher, I called up all the kind of Methodist mafia on route, and we had potluck suppers and things and on the fifth day when we got to the city of Burlington, and there was Bernie on the outskirts of the city to greet us and walk in with us, there were a thousand people walking, which as you all know, for Vermont, is a lot of people.

It was a good day, and we got all of our political leaders to agree to work on climate change. The hard part of that day was picking up the *Free Press* the next morning and reading a story that said that those thousand people were the largest demonstration that had yet taken place in the United States about climate change.

When I saw that, I understood why we were losing. It's because we had assumed that reason, alone, would be enough to carry this fight. That if you got the best scientists in the world, and they went up to Capitol Hill, year after year, and said the worst thing is happening, eventually we'd do what was obvious: make the switch to renewables, put a price on carbon, all the things that people have been talking about.

But that didn't happen because reason is not what decides these things alone. We had the superstructure of a movement: the scientists, the economists, the policy people. The only part of the movement we didn't have, was the movement part; something there to give it some heft, to make people pay attention. So that's what we deliberately set out to do, try, and build as we started 350.org.

When I say *we*, we started it with myself and seven undergraduates in Middlebury College in 2008. Five years ago just about exactly. They were game but we didn't have any money or anything so it was kind of ludicrous. There were seven of them, there were seven continents, each one took one. We went to work.

We spread out around the world, and about a year later we had our first big global day of action. We did not know; we'd asked people to do something that day, wherever they were, to kind of raise the alarm, we didn't know how well it would go. No one had ever really tried anything like this before.

We got the first sense that it might work about 36 hours early. We were sitting around this little borrowed office, and the phone rang and it was our leader in Addis Ababa in Ethiopia, organizing Ethiopia for us.

So, she's on the phone, she's our leader, like most of the world, and this is a good point for the high school people here, she's 17 years old. She was running Ethiopia for us, a volunteer like all of us.

And she was in tears. She said "We can't do our demonstration on Saturday, the government just took away our permit. So we're doing it today before they can stop us." Which was brave, but that's not why she was in tears.

She was just like: "We so wanted to do this at the same time as everyone else in the world, we don't want to spoil it for everybody, we want to be part of the big thing." She said, "But, we do have fifteen thousand young people right now out in the street in Addis Ababa." I said "Don't worry about the date. It's okay."

And for the next 48 hours, the pictures just streamed in. In fact the next one came from US troops in Afghanistan. Parked our Humvee, made a 350 with sand-bags, we're not going to drive this week, we're going to walk.

But most of the pictures, and this was important for me to see, because I had spent my whole life hearing that environmentalism was kind of something for rich white people, and if you had to worry where your next meal was coming from you wouldn't be an environmentalist and on and on and on. It just turned out not to be true. Most of the people that we were working with around the world, were poor, and black, and brown, and Asian, and young, because that's what most of world is made up of.

And the pictures just rolled in. They didn't have a digital camera, so they just took a picture, tried to develop it in the home darkroom, it didn't come out so well so they had to write in what their banner had said. But there they were, part of it.

Huge involvement by religious leaders. That's the head of Muslim South Africa, of indigenous traditions, the Anglican archbishop of South Africa, at the head of a huge march. There's our biggest evangelical college.

I've been to Bethlehem to do some organizing. A hard place to get to, but the Dead Sea is shrinking rapidly. Too many military barriers in the way, so the Jordani-ans said we'll make the big "3" on our beach, the Palestinians said we'll do the "5," and the Israelis said we'll take care of the "0," and it was a glorious day.

And across China, and those are our friends in the Maldives. That's a country… you know, you think we have it tough with Irene, there's a kind of permanent Hur-ricane Irene going on there. The highest point in the country is a meter and a half above sea level. That country's been there for 5,000 years, but it's probably not going to make 5,100, though they're fighting as hard as they can.

People who didn't look like people thought environmentalists look like. Every woman in that "0" there is in a full black burqa, you know? So, they don't look to us like members of the Sierra Club or something, but their hearts are in exactly the same place. They're thinking about the future, not about themselves, you know, do-ing what they can. Even in places, you know, there's this oil-rich sheikdom of Abu Dhabi, you can see some oil-rich sheiks down front. But they're somewhat brighter than the average oil tycoon, 'cause there behind them is the largest solar array on planet Earth. These guys are thinking they might stay rich no matter what, alright, as sort of their plan I think.

There were three or four hundred pictures that ended up from that first day in a file just marked 350. And they were adorable, but they were also hard to look at. The eyes of that girl in the middle just kill me every time I show that picture, because she's probably going to be a refugee before her life is out, and for nothing she's done.

We've gone on to do 15,000 of these demonstrations all over the world, 350 re-ally became the way to kind of educate much of the planet about climate change in places where people didn't know about it, including places in this country. And it's been beautiful, fun, unforgettable, powerful work, that's really meant a lot to us, and some of it's even been this sort of huge art. There's 3,000 people making an image, those of you that remember your history, making an image of King Kanoot trying to hold back the rising ocean. So many people in that image from the Dominican Republic that we had to get a satellite to take a picture of it from outer space to sort of show how big it was.

It's all powerful, and it's all not enough. Along with that education, we've also in the last couple of years had to figure out how to really start taking action. A lot of you have come and helped, and I'm so grateful for it. When we did this big civil disobedience action against the Keystone Pipeline, which turned into the biggest thing like it in thirty years about any issue in this country, twelve hundred and some people arrested. The thing that meant the most to me, was that halfway through it, the two week stint, halfway through it, Hurricane Irene hit, and I can't tell you how badly I wanted to come home and help shovel out and things, but we had work there, and two days later a bus showed up from Vermont, and people telling sort of the stories of how they had this epic trek to find enough bridges and things to get them out of the state and down to Washington. But someone said as they got off the bus, they just said it's too late to stop this Hurricane Irene, but maybe we can do something about the next one and that's why we're here.

And it is why we're here, and it's why that fight has gone all over the country… and why for eighteen months we've been able to hold them at bay with this huge pipeline up to the Tar Sands, and it's very hard, and I have no idea if we're going to be able to win in the end, but we are going to fight it until the very last.

But we're also going to do a lot of other kinds of fighting too, because important as it is to play defense, as important as it is to stop pipelines, and stop coal mines, and wonderful as it was when Vermont became the first in the Union to ban fracking last year, you know…

We're not just going to play defense, we also have to play offense. We also have to get in the face of these guys as much as we can. Partly, it means we have to do all the work of building the renewable energy system that undercuts the business of the fossil fuel industry. That's what Bernie's bill helps do a lot. That's what people in Vermont are doing, Vermont Energy Independence Day, Second Edition is March 21st, 4 PM at the State House. In fact, I think you can now…I think they might be showing the movie they've made over the last year, about all the kind of efforts of Vermont energy independence, which is pretty remarkable.

But it also means really going after the fossil fuel industry directly, as directly as we can. And that's why I'm so glad to see those students here from UVM and from Green Mountain College and from Middlebury and from lots of the other places in this state, where young people are taking the lead.

I want to just finish by telling you about this stuff, this fun we've had in the last year or so. I wrote a piece last summer for *Rolling Stone* magazine. For those of you who keep your back issues, it was the one with Justin Bieber on the cover.

But the strange thing about it was, I got a call the next day from the editor saying, "This is weird, but your piece has ten times more likes on Facebook's than Justin's."

Some of that due no doubt to my chiseled good looks, but more to the fact it just laid out what we know about climate change, this new math, and all you need is three numbers to understand the predicament we're in.

Two degrees is the most that all the world's governments have agreed we can warm and have any chance of kind of surviving. Two degrees is the red line. We've already gone to one degree. We're really kind of down to go to two, but we're going to go pretty near there. We can't go past it. So that's the first number.

Second number: five hundred sixty-five gigatons of carbon. That's how much we could burn, the physicists say, and still have some hope of staying below that. At the current rate of burning of fossil fuel, that will take us fifteen years to blow through, so that's a scary number. It tells you how fast we got to change.

> *And it is why we're here, and it's why that fight has gone all over the country... and why for eighteen months we've been able to hold them at bay with this huge pipeline up to the Tar Sands, and it's very hard, and I have no idea if we're going to be able to win in the end, but we are going to fight it until the very last.*

But the really scary number, and the reason I wrote the piece was the third one. Some financial analysts added up how much carbon all the fossil fuel industries have in their reserves already, stuff they are planning to burn: twenty-seven hundred and ninety five gigatons. Five times as much as the most conservative government and scientist on Earth thinks would be safe.

Once you know those numbers, then you know how this story ends, unless we change the script dramatically. Unless we weaken the power of that fossil fuel industry to do what it wants to do. So we went around the country in November and December. It was kind of fun, we had a bus, we had a bus you could sleep in. We had Johnny Cash's old bus driver. We did 24 cities in 26 nights, in big sold out concert halls. You know, the Orpheum in Boston with three thousand people, and things like that. And what we were trying to do is spark a movement to go after the fossil fuel companies, in particular by getting colleges and then states and cities and religious denominations to divest their money they have invested in these guys. Since, clearly, all they care about is money, we have got to start trying to take some of it away.

And here's the good news. As of today there are two hundred and fifty six college campuses in this country, with active divestment fights now. The *Nation* magazine says it's the largest student movement in decades. There's lots of representatives.

And if you've happened to have gone to a college, you'll know this, that they retain an interest in you, and send you letters and things, you might help out here too, you know?

But what I really want to end by saying, this is such a beautiful room, filled with people of every age, I mean it really shows the diversity. It's very true that we are going to have young people leading this fight, and they have been leading it, and that makes sense because if you are like me and have twenty years left on this planet, well, you know, we'll probably stagger out without the full, worst effects of climate change just breaking over our heads. But if you're twenty right now, and have sixty or seventy years left, this is your life, and it's all your life's going to be by the time you're my age if we don't do something right. It's just going to be staggering from one hurricane to the next, and one wildfire to the next, and civilization will just be a kind of emergency response drill.

That's why we've got to fight, and we've all got to fight. When we did those arrests in Washington, one of the things I said when we sent out the letter was, I do not think that young people should have to carry the whole burden here. Because actually, right now in our economy here, if you're twenty-two, maybe an arrest record is not the absolute best thing for your resume. One of the few good things about growing older is, after a certain point, what the hell are they going to do to you?

And so it was…there's people in this room who came and went to jail, and it was really fun to see how many of them had hairlines more or less like mine. Now we did not ask people, "How old are you?" because that would be rude. But we cleverly said, "Who was President when you were born?" And the two biggest cohorts came from the FDR and Truman Administrations, ok? And on the last day, there was a guy arrested who had a sign around his neck that said "World War II Vet: Handle with Care." He'd been born in the Warren Harding Administration, which is so long ago I'd forgotten there was a Warren Harding Administration.

The only other thing we said was, if you wanted to come get arrested, would you please show up wearing a necktie or a dress. Not because…I mean, I'm a good Vermonter, I just sort of wrap myself in fleece and make it through 'til April. But because we wanted to say something, that really comes out of every one of those pictures I showed you, which is there's nothing radical about what we are talking about here. It's just pure common sense. Radicals work at oil companies. If you're willing to alter the chemical composition of the atmosphere, then you are a deep radical. And if all you are asking for is a world that works a little bit like the world you were born into, than in some deep sense you're kind of a conservative.

So I cannot promise you that we're going to win. There's going to be a million little battles along the way. And some of them we hear in the state. The Senate's going to vote this week on whether to pass this bill that's going to make it really hard to do renewable energy in Vermont, and I hope you call your Senator to ask him not to. But there's going to be lots of battles nationally, and lots of battles internationally. And so on and so forth.

I don't know if we are going to win; we are behind, alright? And the enemy we are fighting isn't even the fossil fuel industry; it's physics, which is a tough enemy to be fighting. But the only thing I want to just assure you of, the only reason I wanted come today, was just to let you know that there is going to be a fight, and you are a huge part of it, and it's going on in every corner of the world. It started in many ways in Vermont and we just cannot thank you enough for all that you are doing to make that fight real and to make it powerful.

Thank you, guys, so much.

# 4

# Wildlife and a Changing Climate

Frans Lanting/DPA/Landov

Northern elephant seal pup (*Mirounga angustirostris*), Ano Nuevo State Reserve, California.

# Plants and Animals in the Balance

All life on the earth exists in a dynamic balance that has existed for millennia. Events such as wildfires, floods, droughts, and seismic activity have always affected the viability of plant and animal species. In the planet's distant past, less common events such as the earth-fall of meteors have had devastating effects on flora and fauna. The most significant such event took place some sixty-five million years ago when an asteroid believed to be roughly the size of Mount Everest struck the area of the Yucatán Peninsula and created the Chicxulub Crater. The energy released by the initial impact and the successive influx of ejecta is believed to have superheated the atmosphere, which in turn triggered a massive firestorm that effectively incinerated at least 95 percent of all living things on the planet.

Whether or not the earth will experience a similar event in the future is hard to predict. Nonetheless, catastrophic events in the planet's past serve to underline the life-and-death relationship that exists between the atmosphere and humankind, animals, and plants. There is a centuries-old parable that says if a frog is placed in cold water and the temperature of the water is slowly raised, the frog will remain in the water and not realize that it is in danger of being boiled to death. However, if the frog is put directly into hot water, it will immediately jump out to save itself. While a catastrophic meteor strike is comparable to the latter event in the parable, the first part of the parable provides a useful analogy for the impact of climate change—specifically, global warming—on humankind. Some scientists argue that humanity remains in a state of blissful ignorance as it continues to risk its own destruction from overheating.

## Climate Change and the Plant Kingdom

Life on earth has always been intimately connected to the stability of the atmosphere. Indeed, the planet's oxygen-rich atmosphere was created by the evolution of plant life in the early oceans. Green plants use the biochemical process of photosynthesis to convert carbon dioxide and water into oxygen and the six-carbon sugar glucose using solar energy. Each molecule of carbon dioxide that undergoes the process results in the formation of a molecule of oxygen. Scientists believe that in the early oceans primitive algae underwent photosynthesis. In photosynthesis, glucose can be "biopolymerized" to produce starch and cellulose, which exist as much larger molecules. These materials came to be used to form the rigid cell walls of plants. As evolution progressed, ocean plants came to colonize dry land, helping to create the variety of terrestrial plants that cover much of the earth's surface today. These plants continue to capture carbon dioxide from the atmosphere and convert it into oxygen and glucose by way of photosynthesis.

When the plants are in darkness, however, the opposing process of respiration dominates. During respiration, plants give off carbon dioxide as the product of their normal metabolic functions. In addition, plant matter that has died—such as fallen deciduous leaves, broken tree limbs, and lost fruit—undergoes the process of oxidative decomposition. Their cellulose and other carboniferous matter react with oxygen in the air and slowly convert back to carbon dioxide, water, and whatever minerals have been trapped during the period in which the plant matter was biochemically alive.

Photosynthetic plants—especially trees—capture carbon dioxide from the atmosphere. In areas such as the rainforests of the Amazon River basin and other tropical areas, so much carbon is captured in the normal photosynthesis/respiration cycle that the level of carbon dioxide in the atmosphere is maintained at an optimum level. In times of high ambient temperatures and low rainfall, plant growth tends to slow, which decreases the amount of carbon dioxide that is captured. Since trees and other cellulosic plants consume carbon dioxide, elevated levels of atmospheric carbon dioxide are beneficial to plant growth, and to a much smaller extent than one might expect, this is true. Yet recent research has revealed that drought in regions of tropical forest have drastically curtailed consumption of carbon dioxide. Instead, trees in affected areas release more carbon dioxide through respiration than they consume. Other studies reveal that when the level of carbon dioxide in the atmosphere passes a certain point, the quantity and quality of plant growth is negatively affected. Tomatoes, for example, become tougher and less juicy, which makes them less desirable as a food product. Other fruits and vegetables react to excessive carbon dioxide in a similar manner.

The amount of carbon dioxide in the atmosphere—and in particular its presence as a by-product of human activity—has long been the principal focus of most discussions about climate change. Carbon dioxide is a known "greenhouse gas." That is, it traps thermal energy in the atmosphere that would otherwise be radiated into space. On average, the amount of carbon dioxide in the atmosphere is small, amounting to approximately 0.3 percent by weight. This amount of carbon dioxide is sufficient to maintain the earth's average surface temperature at a level well above the freezing point of water. Scientists believe it would require only a small increase in the total amount of carbon dioxide present in the atmosphere to produce a significant increase in average global temperature. Increases in global temperatures over recent decades appear to be related to increasing levels of atmospheric carbon dioxide stemming from human industrial activity. The actual causes of climate change are not known for certain, but it is affecting the plant kingdom profoundly. As warmer climes extend farther and farther into traditionally cold zones, a general migration of flora is also taking place. Fir and pine forests are extending to more northerly regions and higher altitudes in mountains as snowcaps and glaciers disappear. Trees and other plants typical of warm regions are also slowly extending their range. That same flora is less able to withstand temperature changes in its native habitat and is becoming a net generator of carbon dioxide instead of a means by which atmospheric carbon dioxide is captured.

In the oceans, where algae and plankton form the base of the food chain, the effects of climate change have been unpredictable at best. Under normal temperatures, massive blooms of algae spring quickly in near-surface ocean waters, fed by nutrients upwelling on cold currents as the oceans' waters circulate. These blooms provide rich feeding grounds for a great variety of fish and other creatures. As climate change progresses, however, there is concern about how such blooms will be affected.

## Climate Change and the Animal Kingdom

The earth's ecological balance is kept in check by the mutual dependency of global food chains. The untold billions of plankton that feed on algal blooms attract large numbers of fish and other creatures that feed on them. These in turn attract other, larger creatures, and so on, up the food chain. Omnivorous humans, the top predators on the planet, depend on this cycle for much of their food supply. Climate change is already bringing about changes in the relationship between humankind and its oceanic food supply. Reports of various fish species now being found well out of their traditional areas occur with increasing frequency, along with reports of species depletion in other areas. Changing water temperatures associated with climate change are thought to be the primary cause of such species relocation throughout the world's oceans.

The most apparent of changes ascribed to climate change in terrestrial areas relate to changes in the ice and snow cover of the Arctic and Antarctic regions. As global temperatures increase, the polar sea ice cover has been observed to be melting at a rate that exceeds the predictions of scientists. As the ice recedes, animals that have evolved and adapted to life involving a large continuous ice cover are forced to cope with and adapt to life with much less ice cover. The polar bear population and the aquatic animals on which they depend for food have become a central focus of scientific research investigating the effects of climate change in the Arctic. A polar bear spends much of its life wandering the Arctic ice pack in the hunt for the seals and walruses on which it normally feeds. On the continuous ice pack, this hunt is uninterrupted. However, as sea ice disappears, the polar bear faces large stretches of open water between ice floes. The hunt for food thus becomes much less productive and more dangerous as the bears are forced to swim long distances through open water.

Global warming is influencing the normal existence of animals in a variety of ways. The pika, a member of the rabbit family, lives exclusively in the cold, high alpine regions of mountain chains. Warmer temperatures have driven pikas to higher altitudes where there is less space for them to subsist. During the warm weather of summer, the pika tends to move down slope to shelter in the cool spaces between fallen rocks. Warmer temperatures down slope mean that this shelter may not exist for the pika, and scientists have observed that pika populations have plummeted in numbers and even disappeared entirely in many alpine areas.

Insects have also been affected by climate change. Since they are predominately herbivores, what affects the flora on which they feed also affects insects. There is an intricate relationship between carbon dioxide levels in the atmosphere, the nature of the phytochemicals produced by plants, and the feeding behavior of insects. With elevated carbon dioxide levels, plants produce more carbohydrate-based phytochemicals at the expense of alkaloids and other nitrogen-based phytochemicals. Insects then adapt to the lower levels of nitrogen-based compounds by consuming more of the plant matter in order to acquire the nitrogen necessary for their own biochemical metabolism. More food for insects, coupled with warmer temperatures that encourage insect reproduction, inevitably leads to more insects and hence more crop depredation. More insects results in more food for insectivorous birds and other animals, which allows for imbalances in other realms of the food chain. Thus far, scientists have established that the effects of global warming throughout the animal kingdom are unpredictable and generally negative.

### Effects on Plants and Animals as Indicators of Climate Change

While scientists agree that the amount of carbon dioxide in the atmosphere has increased and that temperatures are increasing worldwide, there remains some debate about the specific causes of climate change. Most scientists agree that global warming has been caused by the continuous emission of carbon dioxide into the atmosphere that is a by-product of human industrial activity—specifically, the burning of fossil fuels such as coal, gas, and oil. Some scientists argue that regardless of foreign influences, climate change is part of a natural process that humans simply do not yet fully understand. Nevertheless, it is clearly understood that climate change will continue to affect both the plant and animal kingdoms in numerous ways in the future.

# Can Evolution Beat Climate Change?

By David Biello
*Scientific American,* April 15, 2013

The oceanic pincushion known as the purple sea urchin relies on its many spines and pincers for protection and food. An inability to form its spiny shell would devastate the species, which thrives on rocky shores off North America's west coast. Unfortunately for the purple sea urchin, higher carbon dioxide levels in the atmosphere as a result of human fossil-fuel burning presage a more acidic ocean that might make it harder to form such shells.

But new research suggests that the purple sea urchin may have the genetic reserves to combat this insidious threat. A study published in *Proceedings of the National Academy of Sciences* on April 8 found that exposing purple sea urchins to the kinds of acidified ocean conditions possible in the future unleashed genetic changes that may help the animal survive. The researchers showed that, although the exterior of sea urchin larvae changed very little, their genetics adapted to high $CO_2$ environmental conditions in a single life span.

Shifting environmental conditions have always played an outsize role in driving evolution. A climate change from cold to hot transforms everything an organism needs to survive and thrive, so each animal, plant, microbe and fungus species must adapt or die— as happened during the transition out of the most recent ice age. So the question isn't if the current bout of human-induced climate change will drive evolution, but how—and maybe when?

In the case of the purple sea urchin exposing urchin larvae to current and projected levels of ocean acidification—and then sampling their genes at set dates of development—revealed a population undergoing genetic changes under more acidic conditions. Simply put, those larvae with versions of genes better adapted to high $CO_2$ conditions became more common. "In a sense, it is the beginning of evolution," explains biologist Melissa Pespeni of Indiana University Bloomington, who lead the experiment. "Only the individuals with the 'right' gene copies would be able to pass their genes on to the next generation."

> *This research suggests that the key to any evolved response to ocean acidification is having enough diversity in the population to allow natural forces to pick and choose what survives and thrives.*

The genes in question code for proteins involved in processes like extracting shell-building minerals from seawater or fat metabolism. The larvae exposed to today's conditions showed none of the changes seen in those exposed to higher $CO_2$ conditions. And the effect grew over time—some selection could be detected after one day, but an even more prominent shift was apparent by the seventh day of development.

Previous studies have suggested that such purple sea urchins—and other shell-forming organisms—would struggle to grow and develop as the ocean grew more acidic, results that the new study ascribes to differing lab conditions, particularly how densely the urchin larvae are packed. Although purple sea urchins like to cluster close together, testing larvae under these conditions may have exacerbated the impact of ocean acidification.

Of course, purple sea urchins are unlikely to face stress only from ocean acidification; other threats include overfishing for urchin roe. This research suggests that the key to any evolved response to ocean acidification is having enough diversity in the population to allow natural forces to pick and choose what survives and thrives. Plus, "we don't know if there are negative side effects of such rapid evolutionary change," Pespeni notes. The genes lost as a result of selection to cope with high acidity could prove to play an important role in anything from avoiding predators to immune system responses.

Regardless, such evolution does not have to be slow, as this sea urchin work shows. Research in soil mites published on April 8 in *Ecology Letters* reaffirms that point, finding that laboratory-induced natural selection—in this case for shorter maturation time—can work in as little as 15 generations.

Then again, the purple sea urchin may be uniquely prepared to face a future of increased acidity. The upwelling ocean environment where it lives periodically fluctuates between high- and low-$CO_2$ seawater conditions anyway. That means the population may have retained the genetic capacity to deal with high $CO_2$—Pespeni notes they have more genetic variability than most other organisms, a genetic reservoir that may serve the urchins well as they face the effects of climate change.

That also suggests other organisms at sea and on land without that history of exposure will not share the same genetic resilience—as well as often lacking the purple sea urchins' large population sizes. "Right now, it's really unclear what sorts of species are likely to be able to evolve their way out of trouble," says ecologist Dov Sax of Brown University, who was not involved in this research. "It's a giant question that needs to be resolved and feeds into the issue of who is most at risk of extinction from climate change."

# Elephant Seals Help Find Missing Piece in Global Climate Puzzle

By Richard A. Lovett
*Scientific American,* February 25, 2013

By tracking the voyages of elephant seals off Antarctica, and with the help of satellite imaging and undersea sensors, researchers have discovered a long-elusive source for the deep-ocean streams of cold water that help to regulate the Earth's climate.

Antarctic bottom water (AABW) is cold, highly saline water that forms near the shores of Antarctica. Being denser than typical seawater, it sinks to the depths and then moves north in sluggish currents that spread across the globe.

Three sources of AABW were known until now. The first, in the Weddell Sea, was found in 1940; two others were found in the Ross Sea and along the Adélie Coast of East Antarctica in the 1960s and '70s. But for years, researchers have suggested that these were not the only ones. In particular, water samples from an area called the Weddell Gyre contain atmospheric pollutants known as chlorofluorocarbons (CFCs), indicating that the deep water came into contact with the air far too recently to have been carried there from one of the known AABW sinks.

Now, Kay Ohshima, a physical oceanographer at Hokkaido University in Sapporo, Japan, and his colleagues have traced that water to a fourth AABW source, in the Cape Darnley polynya. Their results are published today in *Nature Geoscience.*

Polynyas are regions of open water near sea ice that are kept from freezing by wind and currents that sweep newly formed ice away. Polynyas have relatively high salinity, because most of the salt in sea water is expelled as it freezes.

Armed with the hypothesis that the missing source might be such a polynya, the researchers used satellite sensors to hunt for polynya regions where ice formed particularly rapidly. When satellite data suggested that Cape Darnley might be a candidate, the researchers moored instruments on the seabed, hoping to spot the descending current. In addition, they relied on

> *The new finding fills a gap in researchers' understanding of the Southern Ocean's role in global climate, "including carbon dioxide, temperature, the stability of the Antarctic ice sheet and changes in sea level," says Richard Alley*

data from elephant seals (*Mirounga leonina*) tagged with instruments that monitor ocean conditions.

"The seals went to an area of the coastline that no ship was ever going to get to, particularly in the middle of winter," says Guy Williams, a physical oceanographer at the Antarctic Climate and Ecosystems Cooperative Research Center in Hobart, Australia, and a co-author of the study.

The elephant seals confirmed the researchers' hunch. "Several of the seals foraged on the continental slope as far down as 1,800 meters," he says, "punching through into a layer of this dense water cascading down to the abyss. They gave us very rare and valuable wintertime measurements of this process."

The new finding fills a gap in researchers' understanding of the Southern Ocean's role in global climate, "including carbon dioxide, temperature, the stability of the Antarctic ice sheet and changes in sea level," says Richard Alley, a geophysicist at Pennsylvania State University in University Park, who was not part of the study.

Still, Williams and Ohshima say that the Cape Darnley polynya represents, at most, about one-eighth of the world's AABW, and that other, similar sources might remain to be discovered.

Michael Meredith, a polar oceanographer at the British Antarctic Survey in Cambridge, UK, who wrote an accompanying commentary on the study, says that if the total rate of AABW formation declines, the resulting changes in cold-water circulation could have important effects on global climate, letting the ocean depths warm and thereby changing the rate of heat exchange between Antarctica and the tropics. Moreover, he says, sea levels could rise—owing to the fact that water expands as it warms—and temperature changes could affect deep-sea ecosystems.

# Why a Hotter World Will Mean More Extinctions

By Bryan Walsh
*Time,* May 13, 2013

The end of last week saw the carbon concentrations in the atmosphere finally pass the 400-part-per-million threshold. That means carbon levels are higher now than they've been for at least 800,000 years, and most likely far longer. There's nothing special per se about 400 parts per million—other than giving all of us a chance to note it in article like this one—but it's a reminder that we are headed very fast into a very uncertain future.

Parts per million and global temperature change, though, are just numbers. What matters is the effect they will have on life—ours, of course, but also everything else that lives on the planet earth. I've written before that while I certainly worry about and fear the impact that unchecked climate change will have on humanity, I also feel relatively—relatively—confident that we will, in some ways, muddle through. Human beings have already proved that they are extremely adaptable, living—with various degrees of success—from the hottest desert to the coldest corner of the Arctic. I don't think a future where temperatures are 4°F or 5°F or 6°F warmer on average will be an optimal one for humanity, to say the least. But I don't think it will be the end of our species either. (I've always favored asteroids for that.)

But the plants and animals that share this planet with us are a different story. Even before climate change has really kicked in, human expansion had led to the destruction of habitat on land and in the sea, as we crowd out other species. By some estimates we're already in the midst of the sixth great extinction wave, one that's largely human caused, with extinction rates that are 1,000 to 10,000 times higher than the background rate of species loss.

So what will happen to those plants and animals if and when the climate really starts warming? According to a new study in *Nature Climate Change*, the answer is pretty simple: they will run out of habitable space, and many of them will die.

The Intergovernmental Panel on Climate Change (IPCC) has estimated that 20 percent to 30 percent of species would be at increasingly high risk of extinction if global temperatures rise more than 2°C to 3°C above preindustrial levels. Given that temperatures have already gone up by nearly 1°C, and carbon continues to pile up in the atmosphere, that amount of warming is almost a certainty.

But Rachel Warren and her colleagues at the University of East Anglia (UEA), in England, wanted to know more precisely how that extinction risk intensifies with

> **The Intergovernmental Panel on Climate Change (IPCC) has estimated that 20 percent to 30 percent of species would be at increasingly high risk of extinction if global temperatures rise more than 2°C to 3°C above preindustrial levels.**

warming—and whether we might be able to save some species by mitigating climate change. In the *Nature Climate Change* paper, they found that almost two-thirds of common plants and half of animals could lose more than half their climatic range by 2080 if global warming continues unchecked, with temperatures increasing 4°C above preindustrial levels by the end of the century. Unsurprisingly, the biggest effects will be felt near the equator, in areas like Central America, Sub-Saharan Africa, the Amazon and Australia, but biodiversity will suffer across the board.

In statement, Warren said:

> Our research predicts that climate change will greatly reduce the diversity of even very common species found in most parts of the world. This loss of global-scale biodiversity would significantly impoverish the biosphere and the ecosystem services it provides.
>
> We looked at the effect of rising global temperatures, but other symptoms of climate change such as extreme weather events, pests, and diseases mean that our estimates are probably conservative. Animals in particular may decline more as our predictions will be compounded by a loss of food from plants.

The good news is that a much hotter future isn't a certainty. If global greenhouse-gas emissions peak within the next few years and begin to decline afterward, the UEA researchers suggest that we can preserve many species that would otherwise be lost. Even if the peak is delayed until 2030, fewer species will go extinct. Mitigation will also buy us time to figure out adaptation strategies for some species that are being displaced by climate change.

Of course, there's virtually no chance that global emissions will peak within the next few years—and odds aren't much better even if we give ourselves another 15 years. Even if we can curb warming, an expanding and (hopefully) richer human population is going to put more and more pressure on what we once quaintly called the natural world. The future is going to be difficult for a lot of nonhumans—and tough for a lot of humans too. But a crowded world is a better bet than a hot and crowded one.

# Global Warming's Evil Twin Threatens West Coast Fishing Grounds

By Pete Spotts
*Christian Science Monitor,* June 14, 2012

Over the next few decades, coastal waters off of California, Oregon, and Washington are in danger of becoming acidic enough to harm the rich fisheries and diverse marine ecosystems there, according to a new study. Blame it on global warming's evil twin.

The process changing the seas' chemistry has been dubbed "ocean acidification." It refers to the impact that rising carbon dioxide levels in the atmosphere are having on seawater. $CO_2$ levels are increasing as humans burn fossil fuel and change land-use patterns. The oceans absorb up to 26 percent of those emissions—a number that is expected to go up as the Arctic Ocean loses more of its summer sea-ice cover.

By 2050, the team conducting the study estimates, more than half the near-shore waters governed by the California Current system are likely to become so acidic throughout the year that many shell-building organisms will be unable to maintain their armor. That point could come within the next 20 to 30 years for some sea-floor habitats on the continental shelf, the researchers estimate.

While the team anticipated it would see marine conditions deteriorate with rising atmospheric $CO_2$ levels, "I was really surprised to see how quickly some of these changes will be occurring," says Nicolas Gruber, a biogeochemist at the Swiss Institute of Technology in Zurich who led the team.

The team "points out fairly clearly that if it wasn't for anthropogenic carbon, we wouldn't be passing that tipping point" from encroaching, acidic water, says Richard Feely, a senior scientist at the National Oceanic and Atmospheric Administration's Pacific Marine Environmental Laboratory in Seattle. "That's a very important part of that paper."

Although the study doesn't directly address the question of which creatures get hit hardest first, the team does suggest that other studies indicate oysters could be vulnerable, especially as juveniles. Still, the team acknowledges that some organisms are hurt by even small changes in acidity, while others can tolerate larger changes, at least for relatively short periods of time.

The results were posted Thursday on *ScienceExpress*, the online outlet for the research journal *Science*. *Science* will publish the results in paper form later.

## A Delicate Environment

To some, the phrase "ocean acidification" may trigger visions of house keys melting in the surf. While the changes are more subtle than that, at least on a human scale, they can be harmful to many forms of marine life.

As $CO_2$ dissolves in the ocean, seawater gradually acidifies. Shell-building marine creatures—ranging from tiny plankton to headliners for bouillabaisse and bisque—have a far more difficult time building and maintaining their protective shells. The tiny creatures that build coral reefs also have a harder time drawing on the chemical construction materials once available to them.

In this new research, Dr. Gruber's team conducted modeling studies of the effect that rising $CO_2$ levels are likely to have on ocean chemistry along a stretch of coastline influenced by the California Current system, which runs along the West Coast from British Columbia through the southern end of Baja California. The area the team focused on stretches from Point Conception, near Santa Barbara, northward to the California-Oregon border.

The team selected the location because of its high biodiversity and its economic value as a source of seafood. But they also selected it because the waters are naturally more acidic than waters in many other parts of the Pacific.

The reason: Water rises from the deep ocean during natural upwelling events, and it's naturally more acidic than water near the surface. The upwelling brings nutrients that have led to the area's high biological productivity. But the water's relatively high acidity means shell-builders start their process in already stressful conditions. This is particularly true of juveniles.

Over time, organisms have adapted to this "background" acidity. From a shell-builder's perspective, as long as the seawater is overstocked with calcium, magnesium, and dissolved carbonates, it can process these to build and maintain its shell and do so while spending a minimum amount of energy on the job. From a biogeo-chemist's perspective, water is supersaturated with the necessary building blocks.

## Ocean Acidification's Effect

But the ocean also is absorbing roughly 26 percent of the carbon dioxide humans are releasing to the atmosphere through burning fossil fuel and through land-use changes. This $CO_2$ from above dissolves in the seawater, forming a very weak carbonic acid.

As seawater becomes increasingly acidic, it eats into the water's inventory of calcium and magnesium. If the seawater becomes undersaturated in these minerals, shells dissolve because their inhabitants can't maintain them in the face of insufficient raw materials.

Generally, the water itself isn't lethal, Gruber says. Instead, it retards the organism's growth, making it harder to survive.

Even with relatively low growth in human-triggered $CO_2$ emissions, by 2050 the saturation levels of key minerals drops quickly. Within the next 30 years, the top 200 feet of water near shore is likely to become undersaturated throughout the summer. By 2050, more than half the waters in the study area become undersaturated all year. But sea-floor habitats could see year-long undersaturation within the next two to three decades, the study projects.

## A Double Whammy

The results come against a backdrop of increasing acidification in waters throughout the Pacific Basin. In a study accepted for publication in the journal *Biogeochemical Cycles*, Dr. Feely and colleagues show that on average, what he terms the "corrosive layer" of ocean water has risen through the upper 2,000 feet of water at a pace of about three to six feet per year between 1991 and 2005.

*As seawater becomes increasingly acidic, it eats into the water's inventory of calcium and magnesium. If the seawater becomes undersaturated in these minerals, shells dissolve because their inhabitants can't maintain them in the face of insufficient raw materials.*

"But there are locations where it rises much faster than that," he continues. "The Washington-Oregon-California coast is particularly vulnerable to these kinds of processes because of the combined impacts of anthropogenic $CO_2$ and upwelling. This draws highly corrosive waters onto the [continental] shelf."

Although the California Current system was the focus in the new study from Gruber and his colleagues, the problem occurs elsewhere. Other key locations include a vast region of upwelling stretching north off the coast of Peru, as well as off Portugal and in two locations along Africa's west coast. In the Atlantic, however, the effects of acidification are less pronounced because it harbors a larger inventory of calcium, magnesium, and carbonates than does the Pacific, Gruber says.

# Common Plants, Animals Threatened by Climate Change, Study Says

By Neela Banerjee
*Los Angeles Times,* May 12, 2013

Climate change could lead to the widespread loss of common plants and animals around the world, according to a new study released Sunday in the journal *Nature Climate Change.*

The study's authors looked at 50,000 common species. They found that more than half the plants and about a third of the animals could lose about 50 percent of their range by 2080 if the world continues its current course of rising greenhouse gas emissions.

Climate change affects the availability of nutrition and water for animals and plants. The narrowing of the geographic range of different common species means that plants and animals readily found in a given area could diminish markedly in those areas over the next seven decades.

"This study...tells us that the average plant and animal will experience significant range loss under climate change," said the study's lead author, Rachel Warren, of the Tyndall Centre at the University of East Anglia, United Kingdom.

Warren said that until now, much climate change research had focused on the plight of rare species rather than common animals and plants. The study's conclusions are "entirely consistent with what others are finding around the world," said Peter B. Reich, professor of forest ecology at the University of Minnesota in St. Paul, who read the report.

> *Reich said that in Minnesota, it appears very likely that spruce, fir and aspen forests will retreat north into Canada over the next several decades as the state warms.*

The new study predicted that plants, reptiles and particularly amphibians would face the greatest risks from climate change. It also concluded that sub-Saharan Africa, Central America, the Amazon region and Australia would likely lose the most species of plants and animals. It projected "a major loss of plant species" in North Africa, Central Asia and South America.

A map of a species' habitat can be thought of as a paint-by-numbers picture, Reich explained. If the species now occupies a certain number of spaces painted red, there will be fewer red spaces—less habitable terrain—70 years from now, he said.

According to Jeff Price, a coauthor of the study and visiting fellow at the Tyndall Centre, "coffee, chocolate, teak, sugar maple, pineapple and cotton (at least some of the major types) all show large contractions in their climatic ranges under the baseline climate change scenario."

The climactic range is the habitat where species not only exist but can thrive in numbers amid competitors.

Reich said that in Minnesota, it appears very likely that spruce, fir and aspen forests will retreat north into Canada over the next several decades as the state warms.

While the study looked at the effects of climate change on species' geographic range, it made conservative estimates about how global warming could stoke diseases, pests or natural disasters that would affect species, Warren said.

Warren said, "Animals in particular may decline more as our predictions will be compounded by a loss of food from plants."

The study concludes that such widespread losses could be avoided if the global community moved swiftly to reduce emissions of heat-trapping greenhouse gases that cause climate change.

Immediate, substantial steps could reduce species losses by 60 percent and give plants and animals another 40 years to adapt, the study predicted. Slowing or cutting greenhouse gas emissions would help the Earth avoid the higher temperatures and subsequent habitat changes that could drive species loss, the study said.

Chances for averting higher temperatures, however, appear remote. On Friday, the National Oceanic and Atmospheric Administration reported that the ratio of carbon dioxide in the atmosphere has surpassed 400 parts per million in an average daily reading at Hawaii's Mauna Loa Observatory, the highest concentration of the greenhouse gas in human history.

Climate scientists have calculated that the world needs to keep carbon dioxide emissions from crossing the 400 parts per million threshold in order to avoid a rise of 2 degrees Celsius (3.6 degrees F) above the average global temperature of pre-industrial times and profound changes to nearly every aspect of life, including species loss.

# The Climate Change Conundrum: What the Future Is Beginning to Look Like

By Jordan Carlton Schaul and Michael Hutchins
*National Geographic,* January 10, 2013

**Jordan:** In collaboration with the National Wildlife Federation, and Arizona State University scientists, the United States Geological Survey (USGS) recently published a technical report as a contribution to the 2013 Third National Climate Assessment. Entitled the "Impacts of Climate Change on Biodiversity, Ecosystems, and Ecosystem Services," the report examines changes in the geographic range of species and the timing of significant life history events for these respective animal species. According to the authors, these dramatic developing ecosystem dynamics are mixing species together that have not previously interacted and are creating mismatches between animals and their food sources. Can you elaborate on these points?

**Michael:** Yes, one of the most pervasive and problematic impacts of climate change on wildlife will be its influence on plant and animal phenology—changes in seasonal life-cycle events like blooming or migrations, that take place as a result of the changing weather patterns (Inkly, D., Stault, A., and Duda, M.D. 2009. Imagining the future: Humans, wildlife and global climate change. pp. 57–72 in Manfredo, M.J. et al. [eds.] *Wildlife and Society.* Washington, DC: Island Press). Rising temperatures are resulting in earlier springs and longer growing seasons and this has a variety of different environmental influences, including, but not limited to, melting ice packs and glaciers, less snowfall and runoff, changes in the timing of plant and insect emergence, and alterations in animal behavior. A good example is the mistiming of the arrival of migratory birds into temperate regions. Reproduction is generally timed to coincide with an abundance of food; however, if the birds arrive or begin their nesting and breeding cycles too early or too late, sufficient food may not be available to meet the increased energetic demands of reproduction. The same principle is true of emerging butterflies and moths. If butterflies and moths leave their cocoons too early or too late, that is, before or after food or host plants have emerged and bloomed, the mistiming could be fatal or interfere with reproduction. Two winters ago, I had cabbage butterflies flying in my yard in mid-December after a few days of unusually warm weather in the Washington, DC, metropolitan area. Warmer temperatures are also impacting migratory behavior and geographic ranges of butterflies, many of which are expanding their ranges into more northern regions.

Changes in habitat as a result of changing climatic conditions (wetter, dryer, warmer, colder) will also have significant effects on wildlife and their habitats. Warming temperatures and shorter cold spells, for example, are creating ideal conditions for insect pests, such as bark beetles, which are rapidly changing forest composition. Under a very moderate 2°C warming scenario, for example, the mountain pine beetle is likely to seriously threaten the Rocky Mountain white bark pines, which provide food for many wildlife species. Extreme weather, such as high winds, tornados and hurricanes, can also cause extensive blow downs, which can subsequently result in rapid changes in forest composition. Climatic changes are gradually replacing some species with others. Northern hardwood forests dominated by silver maple, swamp white oak, and green ash stands are being replaced by oak, pine, and red maple as their ranges shift northward. Another example is the loss of sugar maple trees, which is affecting the maple syrup industry. In areas that are becoming warmer and dryer, the incidence of forest fire is also increasing, thus multiplying risks to wildlife communities. As you note, rising sea levels from the melting of the polar ice caps and glaciers has the potential to flood low-lying islands and coastal areas, thus resulting in a loss of important wildlife habitats. Disease could also be a problem, as a result of warming temperatures. In particular, disease vectors, such as mosquitoes and ticks, may become more abundant or emerge earlier than normal, thus increasing the risk of transmission (Kutz, S., Shock, D., Brook, R., and Hoberg, E. 2008. Impending ills: Impacts of climate change on infectious diseases in wildlife. *The Wildlife Professional* 2(3): 42–46).

Last, but not least, some invasive species could also be favored by climate change, many of which are already significant threats to our native fauna and flora. In fact, as I mentioned in our previous interview, climate change may result in our having to redefine the concept of species invasions. Changing environmental conditions may result in a mixing of indigenous species that do not normally co-exist, except perhaps at the periphery of their ranges. This could, among other things, result in increased competition.

**Jordan:** The polar bear has emerged as an iconic indicator species of climate change, but there are so many other species that the general public is wholly unaware of with regard to climate change and its impact on entire populations of extant fauna. Do you agree? What are some more obscure species that may be affected by climate change and what are some of the cascading effects that these changes will have on ecosystems?

**Michael:** Absolutely. Polar bears, as an iconic species, have gotten a great deal of attention in the popular media. However, there are many lesser-known species, particularly those that live in polar or alpine habitats that are being affected by climate change. It is estimated that one tenth of the Western Hemisphere's mammals may not be able to "outrun" climate change by moving to more suitable habitats. Many of these are species that the public may have little awareness of. For example, the American pika, a small lagomorph (related to rabbits) which lives on talus slopes in alpine and subalpine habitats, is being heavily impacted by climate change. Local extinctions have increased ten-fold in the last decade and biologists have observed that the animals are moving to higher elevations as temperatures have increased (Beever, E., and Wilkening, L. 2011. Playing by new rules: Altered climates are affecting some pikas dramatically. *The Wildlife Professional* 5(3): 38–41). Similarly, four subspecies of bearded seals and two populations of ringed seals have recently been listed as threatened or endangered under the Endangered Species Act, primarily due to the impacts of climate change. The little-known white lemuroid opossum, a rare marsupial found in northeastern Queensland, may be the first mammalian victim of climate change. The animal has not been sighted in three years, and it is thought to be highly sensitive to temperature change. Just four or five hours of temperatures over 30 degrees could wipe out the highly vulnerable species, as under extreme heat, they are unable to maintain their body temperature. Many invertebrates, like butterflies, whose entire life-cycle is impacted by climate change, receive little attention, yet are essential plant pollinators and food for birds and other animals. Scientists are expecting local extinctions of a third of butterfly species in response to climate change. Even less attention is being given to the smallest of organisms—the plankton at the bottom of the food chains in the world's oceans or the terrestrial microbes that break down dead organisms in terrestrial environments and recycle their bodies back into the earth. Yet these too are being impacted by climate change. The impacts, however, are poorly understood and changing temperatures could produce both positive and negative feedback loops. For example, warming is predicted to increase rates of decomposition in northern latitudes, which could have pervasive effects on nutrient cycles. That being said, these organisms can also be carbon sinks, which could help to sequester atmospheric carbon.

**Jordan:** Certainly some species impacted by climate change will influence a cascade of events impacting ecosystems, which may include further decimation of wildlife populations. These effects on population sizes may be even more dramatic than we see in polar regions because of disparities in biodiversity in polar, temperate and tropical regions of the world. Can you elaborate on this concern?

**Michael:** Yes, this is a legitimate concern. As we know from basic ecology, organisms in functioning ecosystems are interdependent and linked together in complex webs. The loss of one species can therefore result in a cascade of extinctions. Well-known biologists Paul and Anne Ehrlich once likened this to taking the bolts out of a flying aircraft one at a time. It may hold together for a while, but eventually, a wing will fall off and the entire plane will crash (Ehrlich, P., and Ehrlich, A. 1981. *Extinction: The Causes and Consequences of the Disappearance of Species.* New York: Random House). This is especially true for so-called keystone species that are at the center of these complex relationships. To the extent that climate change results in species extinctions—either directly or indirectly—we are likely to see such extinction cascades occur, and this could result in a significant loss of biodiversity.

**Jordan:** While working with endangered wood bison in Alaska in an effort to reintroduce this northern subspecies of American bison—the largest terrestrial mammal in the Western Hemisphere—to the Frontier State, we argued that the return of the extirpated bovid to its historic range in Alaska would serve the interests of climate change mitigation. Bringing back bison should not only complete the faunal assemblage of herbivores in the region, through the reintroduction of a key-note grazing ungulate, but it represents (offers) an opportunity to restore habitat to a state conducive to promoting biodiversity, as bison wallows provide microhabitats attractive to an array of aquatic and terrestrial wildlife and plant species. According to the Convention on Biological Diversity, an increase in biodiversity stifles climate change (or promotes climate change mitigation and adaptation). Can you explain this notion or provide some additional examples of how conserving and adequately managing biodiversity assists with climate change mitigation? What are the relationships among climate change mitigation, biodiversity and land degradation?

**Michael:** There is no doubt that maintaining biological diversity should be a major goal in climate change mitigation. Diversity contributes to the resilience of an ecosystem to perturbations and disturbance. Furthermore, ecosystems and their biodiversity have important roles to play in sequestering carbon.

Informed decision-making and careful management of natural ecosystems for mitigation can enhance this role and help to avoid potential negative impacts. As climate change increasingly puts a strain on ecosystems and their animal and plant populations, it is important to find ways to increase their resilience to current and future climate change. In addition, some species may be able to adapt rapidly to climate change through natural selection, or through changes in phenology.

**Jordan:** Catastrophic weather patterns are among the stochastic events that impact wildlife populations. El Nino events reported in the Pacific come to mind. These isolated events can wipe out conservation sensitive populations of wildlife and particularly those at higher trophic levels like pinnipeds and cetaceans. How are these events different from what climate scientists refer to as global warming and what may be the difference with regard to their impact on wildlife populations?

**Michael:** Like tsunamis and volcanic eruptions, catastrophic weather events, such as hurricanes, tornados, flash freezes and ice storms, extreme drought, and flooding, can have devastating effects on wildlife populations and on the plant communities that sustain them. This is particularly true of small, isolated populations of conservation concern. The smaller and less distributed a species is, the higher the risk that a single weather event could push it over the brink of extinction. These isolated events are different from global warming (which refers to the recent rise in average global air and water temperatures), but the increasing intensity and frequency of such events is consistent with what is expected from a warming planet.

**Jordan:** What is climate change adaptation? What can wildlife managers and conservationists do to promote climate change mitigation?

**Michael:** People should not confuse the term "climate change adaptation" with "adaptation" as it is normally used in evolutionary biology. In the latter, adaptation is used to denote genetic, physiological, and anatomical changes in species as they better adapt themselves to changing environmental circumstances through the process of natural selection (differential survival and reproduction). However, climate change adaptation refers to how we (humans) as stewards of our planet are going to respond to climate change in order to conserve as much biodiversity as possible. At this point, some climate change is inevitable. However, there is a difference between adaptation and mitigation. The severity of change could be mitigated by humans. Chief among these responses would be a massive shift from fossil fuels to renewable energy, which could reduce carbon dioxide and other greenhouse gases in our atmosphere. Although the causes of climate change are complex, there is a direct correlation between the rise of greenhouse gases in our atmosphere since the industrial revolution and a warming planet. So there is some hope that we might head off the worst of impacts with a concerted effort at mitigation. However, some would argue that we already have enough greenhouse gases in our atmosphere to give us another 100 years or more of climate change.

Rushing to clean energy is also a potential problem, as wind, solar and other alternative energy sources can harm wildlife (Hutchins, M., and Bies, L. 2010. How green is "green" energy. *Outdoor America* Winter: 16–17; Leitner, P. 2009. The promise and peril of solar power. *The Wildlife Professional* 3(1): 48–53). Such facilities should not be sited in sensitive ecological areas or, in the case of wind, in areas with large bird or bat concentrations. We clearly need cleaner sources of energy, but it would be counterproductive if such alternatives were also bad for our native wildlife and their habitats.

While serving as Executive Director/CEO of the Wildlife Society, I was very active in trying to identify solutions. In 2011, I participated in a series of leadership forums organized by the US Fish and Wildlife Service at the National Conservation Training Center in Shepherdstown, WV. These meetings initiated the development of a national climate adaptation strategy for fish, wildlife and plants. There are many things that could potentially be done to mitigate the impact of climate change on biodiversity, including the acquisition of new conservation lands to replace those lost to rising sea levels and habitat change, the creation of corridors to facilitate the movement of wildlife and plants, active management of current habitats and native species to promote ecosystem resilience to climate-induced stress, and the creation of new protected areas (Abhat, D., and Unger, K. 2008. Reining in the impacts of climate change. *The Wildlife Professional* 2(3): 33–38).

In ecology, resilience is the capacity of an ecosystem to respond to a perturbation or disturbance by resisting damage and recovering quickly. There is some question as to whether such factors can actually be managed and some scientists have therefore argued that we should plan for recovery rather than resilience. In addition, some have argued that our current system of static protected areas will not be as useful for conservation in the face of a changing climate (Hall, H.D., and Ashe, D.M. 2008. Changing as conditions change. *The Wildlife Professional* 2(3): 11–13). For example, under climate change, a protected area established to conserve a particular species may no longer offer the conditions necessary for its survival, or species may move out of protected areas seeking more favorable conditions (e.g., as in the case of vertical migration). Indeed, some protected areas, such as those in coastal areas, may disappear altogether, thus making the current extinction crisis even more intense. Others have argued that protected areas will be important insurance against extinction. However, to fulfill this role, protected areas will need to be in locations that are predicted to escape the brunt of climate change, and they will need to be linked, rather than isolated. Without a strategically distributed network of protected areas, biomes of the future will likely be limited to weedy and disturbance-tolerant generalist species that might not preserve ecosystem function and services.

In order to help with this kind of land-scape level planning, the US Fish and Wild-life Service developed the concept of Land-scape Conservation Cooperatives or LCCs (Austen, D.J. 2011. Landscape Conserva-tion Cooperatives: A science-based network in support of conservation. *The Wildlife Professional* 5(3): 32–37.). These are regionally ecosystem-based cooperatives con-sisting of representatives from federal and state government agencies, conserva-tion organizations, and academia. Such public-private partnerships are intended to provide the necessary expertise and partnerships to plan for ecosystem resiliency and recovery in the face of climate change and other environmental stressors. It is hoped that, through careful planning and management, resiliency can be increased and adaptation to climate change accelerated. Of course, in order to plan for future climate adaptation, one needs some basis for predicting the impacts on climate change on specific locations and communities. Much work is currently being done to translate general weather models to a scale that will be useful for managers and to try to deal with the uncertainty inherent in such models. In order to facilitate and promote the science behind such predictions, the US Geological Survey has devel-oped a National Climate Change and Wildlife Science Center. This is being further augmented by the Department of the Interior's Climate Science Centers. In 2010 and 2011, I helped to organize a series of meetings that led to the creation of these centers at the request of USGS.

In a changing climate, monitoring and better understanding the effects of such changes also become crucial, and much more of this will have to be done in the future (Beard, T.D, O'Malley, R., and Robertson, J. 2011. New research on climate's front line: Understanding climate change impacts on fish and wildlife. *The Wildlife Professional* 5(3): 26–30). The National Climate Change and Wildlife Science Cen-ter and DOI Climate Science Centers should play an increasingly important role in monitoring, as should the LCCs. However, citizen science is also very useful in this regard and many lay people now volunteer their time to assist in these monitor-ing processes. One program I've been closely involved with is the USA National Phenology Network (USA-NPN), a program that uses citizen and professional sci-entists to monitor biological phenomena associated with climate change. The NPN has long maintained a system of monitoring plant phenology, but more recently, they have developed methods for monitoring wildlife as well (Miller-Rushing, A.J. 2010. Wildlife watchers aid climate research. *The Wildlife Professional* 4(2): 58–61). Check out their website and their "Nature's Notebook" Program to see how you

> *Just four or five hours of temperatures over 30 degrees could wipe out the highly vulnerable species, as under ex-treme heat, they are unable to maintain their body temperature.*

can get involved in tracking the effects of climate change on plants and animals in your own backyard.

A key question going forward will be how to better integrate research with management. Often there has been a rift between university or agency scientists who are seeking to understand natural phenomena and resource managers (Sands, J.P., et al. (eds) *Wildlife Science: Connecting Research with Management*. New York: CRC Press). The former are interested in knowledge for knowledge's sake and the latter in how that basic research will be applied to real life problems. In the past, communication between scientists and managers has been poor or non-existent and that must change. Managers must do a better job communicating their informational needs to scientists, scientists must do a better job of translating their findings for managers, and cooperative planning for adaptation will be critical. If this can be accomplished, then the results of research can be more immediately applied to key resource management challenges associated with climate change.

It is promising that the infrastructure described above has been or is being developed to help our nation address climate change mitigation. However, implementation is going to be expensive, and given our current economic situation, it is questionable whether the funds will be available. One of the most significant questions for conservation in coming years is how we will pay for it? (Hutchins, M. 2012. What the future holds? *The Wildlife Professional* 6(3): 83–87; Hutchins, M., Eves, H., and Mittermeier, C.G. 2009. Fueling the conservation engine: Where will the money come from to drive fish and wildlife management and conservation? pp.184–197 in Manfredo, M.J., et al. (eds.), *Wildlife and Society*. Washington, DC: Island Press). One way that this could be done would be to tax carbon emissions, but this solution has been highly controversial and no action has yet been taken.

**Jordan:** Is choosing to protect our natural wildlife heritage for its intrinsic or fundamental value or for the resource value they provide us going to dictate how we develop or implement strategies aimed at promoting climate change mitigation and adaptation?

**Michael**: This is an interesting question. Climate change represents one of the most significant challenges humankind has ever faced, and, as we go forward, there are many ethical and practical questions that will need to be answered. Wildlife managers and conservationists are acknowledging that some species are going to be lost, and like it or not, we may need to resort to a triage-like decision-making process to decide which species we will focus our attention on and which we will not. But how we will make such decisions and on what basis? As one example, fisheries managers are concerned about the future survival of native western trout as a result of climate change, but they doubt that all varieties can be saved and have suggested that a triage approach might save some. Furthermore, some groups of animals are likely to be more susceptible to climate change than others. Should we focus more attention on particularly vulnerable species or should we write them off and focus our limited resources on those that have a higher probability of survival? The future of life on our planet will depend on how such difficult and complex questions are resolved.

# Dynamics of Coral Reef Recovery

By Beth Polidoro and Kent Carpenter
*Science Magazine,* April 5, 2013

Among all marine habitats, coral reef ecosystems support the highest concentration of marine biodiversity. Yet corals are declining around the world at an alarming rate, mainly as a result of more frequent, larger, and longer-lasting bleaching events observed in recent decades.[1–3] The study of reef resilience and recovery at the isolated Scott Reef, reported by Gilmour *et al.*[4], offers hope that coral reef ecosystems can recover after mass bleaching events, if anthropogenic threats can be greatly reduced or eliminated.

Coral reefs bleach when coral polyps lose their symbiotic zooxanthellae, or microalgae, which are critical for providing nutrients through photosynthesis. Coral bleaching is a response to stress, which can be caused by a number of local or regional anthropogenic disturbances, including sedimentation, pollution, destructive fishing techniques, and overexploitation of fish and other reef inhabitants that maintain optimal reef conditions and macroalgal growth. The increased duration and frequency of global mass bleaching events are also strongly linked to increased sea surface temperatures and ocean acidification associated with global climate change.[5, 6] One-third of the world's 845 species of reef-building corals are considered to be at elevated risk of extinction[7] as a result of the combined effects of global climate change and local anthropogenic impacts. The most pessimistic scenarios estimate that the world's reefs may entirely disappear within this century, but other scenarios based on variable zooxanthellate thermal tolerances and reduced greenhouse gas emissions predict persistence into 2100.[5, 8]

Coral reef recovery is mainly thought to depend on recruitment or arrival of larvae from distant, interconnected reef ecosystems. Local recruitment from surviving individuals, such as those in deeper or adjacent areas, is thought to occur for various marine populations such as marine fishes,[9, 10] but has not been well documented for coral species. Estimates of maximum dispersal distance and larval survival for reef-building coral species are sparse. However, evidence that the larvae of five reef-building coral species survive longer than expected suggests that long-distance dispersal, although rare, may be greater than currently assumed and highly variable depending on the coral species.[11]

The relatively rapid recovery of coral cover following a mass bleaching event at the isolated Scott Reef documented by Gilmour *et al.* suggests that corals can recruit from local sources, especially in the absence of anthropogenic disturbances,

> *The increased duration and frequency of global mass bleaching events are also strongly linked to increased sea surface temperatures and ocean acidification associated with global climate change.*

which have been shown to slow down recovery.[12] Continued anthropogenic pressure can cause changes in ecosystem dynamics such as increased algal growth from pollution and loss of herbivorous fish from overfishing or other impacts. As a result, degraded or destroyed reefs become harder to reseed from healthy, interconnected reefs.[13] Loss of live coral cover can also create negative feedbacks that inhibit ecosystem recovery. For example, loss of structural heterogeneity can cause a decline in the abundance and diversity of reef fishes, especially in isolated reef ecosystems.[10, 14] A recent study in the Caribbean suggests that when the area of live coral in a reef drops below 10 percent, erosion may begin to outpace the growth of new reef structures.[15]

Some coral and fish species are likely to recolonize faster than others, whereas others might not recolonize at all.[10] However, most studies on coral reef loss and recovery are based either on the percentage of live coral cover lost or remaining, or on the loss or recovery of selected genera or key reef species, such as the iconic Elkhorn coral (*Acropora palmata*) and Staghorn coral (*Acropora cervicornis*) in the Caribbean. It is nearly impossible to quantify all coral reef species present either before or after a mass bleaching event, especially in regions such as the Coral Triangle in the Indo-Malay Philippine Archipelago, where more than 500 species of reef-building corals can be found in the same area. Many coral species are very difficult to identify to the species level in the field. Additionally, further taxonomic and/or genetic studies are needed to determine the validity of a large number of reef-building coral species, especially those in the family Acroporidae.[7]

The study by Gilmour *et al.* prompts us to ask how many of the world's reef systems are as isolated from anthropogenic impacts as Scott Reef, and to what degree they are benefiting from local recruitment. What will be the impact of increased ocean temperature and acidification on the long-term survival of all reefs, especially as even isolated reefs are susceptible to climate-driven reef degradation?[14] Across all coral reef ecosystems, more research is needed on which species are more resilient to local or global threats and which species are at highest risk of little or no recovery.

Without more comprehensive species-specific information and research, it remains unclear whether coral species composition and regeneration in recovered reefs can ever reach the same state of previously healthy, undisturbed reefs. It is clear, however, that substantially reducing anthropogenic impacts on coral reefs might at least buy us, and coral reef ecosystems, more time to answer these questions.

## References and Notes

1.  T. P. Hughes *et al.*, *Science* 301, 929 (2003).
2.  C. Wilkinson, Ed., *Status of Coral Reefs of the World: 2008* (Global Coral Reef Monitoring Network and Reef and Rainforest Research Centre, Townsville, Australia, 2008).
3.  O. Hoegh-Guldberg *et al.*, *Science* 318, 1737 (2007).
4.  J. P. Gilmour, L. D. Smith, A. J. Heyward, A. H. Baird, M. S. Pratchett, *Science* 340, 69 (2013).
5.  J. M. Pandolfi, S. R. Connolly, D. J. Marshall, A. L. Cohen, *Science* 333, 418 (2011).
6.  K. R. N. Anthony, D. I. Kline, G. Diaz-Pulido, S. Dove, O. Hoegh-Guldberg, *Proc. Natl. Acad. Sci. U.S.A.* 105, 17442 (2008).
7.  K. E. Carpenter *et al.*, *Science* 321, 560 (2008).
8.  M. L. Baskett, S. D. Gaines, R. M. Nisbet, *Ecol. Appl.* 19, 3 (2009).
9.  R. K. Cowen, K. M. M. Lwiza, S. Sponaugle, C. B. Paris, D. Olson, *Science* 287, 857 (2000).
10. G. P. Jones *et al.*, *Coral Reefs* 28, 307 (2009).
11. E. M. Graham, A. H. Baird, S. R. Connolly, *Coral Reefs* 27, 529 (2008).
12. J. E. Carilli, R. D. Norri, B. A. Black, S. M. Walsh, M. McField, *PLoS ONE* 4, e6324 (2009).
13. T. P. Hughes, *Science* 265, 1547 (1994).
14. N. A. J. Graham *et al.*, *Proc. Natl. Acad. Sci. U.S.A.* 103, 8425 (2006).
15. C. T. Perry *et al.*, *Nat. Commun* 4, 1402 (2013).

# Running Hot and Cold: Are Rainforests Sinks or Tabs for Carbon?

By Sharon Levy
*BioScience,* July 1, 2007

*Conventional wisdom has long held that tropical rainforests act as a sink for carbon dioxide, cleansing the atmosphere of a major greenhouse gas. However, biologists studying the forests of Costa Rica are finding that rising temperatures are causing trees to grow less and to pump out more carbon dioxide, adding to an accelerating pattern of global warming.*

Below me, toucans and green macaws squawk, now and then bursting out of the foliage to show their splendid colors. A half-kilometer away, the topmost branches of a tall tree wave in the still air as a troop of white-faced capuchin monkeys moves through the canopy. I stand at the top of a 42-meter-tall research tower that sprouts out of the old-growth rainforest at La Selva Biological Research Station, in northeastern Costa Rica. On a clear day, I'd have a view of the chain of volcanoes to the west and the bright blue of the Caribbean to the east.

But on this March morning, rain drizzles onto the forest canopy, which surrounds the tower in a rolling panorama of every possible shade of green, then fades into a cloak of low-lying clouds. Up here, where scientists are measuring the exchange of greenhouse gases between the dense trees and the planet's troubled atmosphere, the forest seems to go on forever. In fact, La Selva is only a small haven of surviving rainforest, 1614 hectares surrounded by farmland and other kinds of development. Deborah Clark, who along with her husband and fellow ecologist David Clark has devoted a long career to studying La Selva's complex inner workings, describes it as a "postage stamp" of intact habitat in a landscape heavily altered by the hand of man.

Despite its small size, this forest is having a big impact on the shifting science of climate change and tropical ecology. The Clarks, who started out more than 20 years ago intending to study the basics of tree growth and physiology, have followed their data into an intense debate over the future of tropical forests in a warming world. What they've discovered both fascinates and dismays them.

"When we started out in the early 1980s," explains Clark, "we chose six ecologically different kinds of trees that grow into the canopy. We wanted to figure out, from being babies all the way to becoming old senescent adults, how trees survive, to understand how we can have so many species coexisting." More than

350 different types of trees make up La Selva's forest, which is populated with a spectacular diversity of creatures, including peccaries, agoutis, three kinds of monkeys, and many hundreds of species of birds, bats, and ants. To begin to uncover the secrets of this rich ecosystem, the Clarks and their crew began by simply measuring the diameters of selected trees over and over, year by year. More than a decade into the project, some startling results jumped out at them.

Ecologists had assumed that trees in the consistently warm tropics grew at a slow but steady rate, unvarying from one year to the next. But the trees at La Selva grew less in hotter years, more in cooler ones. Over 16 years of sampling, from 1984 to 2000, dramatic differences in growth rate occurred: In some cooler years the trees added twice as much wood as they did in the scorching El Niño year of 1997–1998. All six species of trees, representatives of divergent plant families with different life histories, showed the same pattern.

## First Hints of Carbon Efflux

Tree growth is an index of the balance between photosynthesis, in which plants fix carbon and release oxygen, and respiration, in which plants use up oxygen and breathe out carbon dioxide. Global temperatures are now rising fast, driven by human emissions of carbon dioxide and other greenhouse gases. The data from La Selva were among the first hints that tropical forests might be pushed by increasing heat to release more carbon dioxide, intensifying global warming. This suggested a reversal of the popular theory that tropical forests act as a sponge, soaking up much of the excess carbon dioxide humans pump into the air.

While grappling with their new findings, the Clarks discovered that their data are consistent with a model of global carbon flux developed by Charles David Keeling, a pioneering chemist with the Scripps Institution. In the 1960s, Keeling, who died in 2005, demonstrated that atmospheric levels of carbon dioxide were steadily rising as a result of human activities. (The two major anthropogenic sources of atmospheric carbon dioxide are combustion of fossil fuels and destruction of tropical forests.) He and his colleagues built their model on measurements of carbon dioxide concentration, taken throughout the 1980s and 1990s at nine stations scattered from the Arctic to the Antarctic, and on estimates of worldwide fossil fuel emissions and prevailing winds.

From this model, Keeling concluded that the amount of carbon dioxide taken up in tropical landmasses peaked in cool years and fell in hotter ones, accounting for year-to-year changes in the amount of human-generated carbon dioxide that stayed in the atmosphere. "The amazing thing," says Clark, "is that the Keeling model, based on global atmospheric gas sampling, is perfectly correlated with our tree growth record here at La Selva. This is the first and only case of a long-term biological data set supporting one of these models. So I think the Keeling model is onto something."

Tropical forests cover a relatively small portion of the Earth's surface, but they are thought to be responsible for more than a third of plant productivity. The La Selva data suggest that trees could be very sensitive to the kinds of temperatures already hitting the tropics. It's not just a question of hypothetical effects as the climate warms; the rainforest has already been panting in the heat, most notably in the intense El Niño year of 1997–1998. A newly published study of forests in Panama and Malaysia shows a temperature-related decrease in tree growth rates that parallels the La Selva findings, and recent studies of rice fields in the tropics suggest that crop productivity drops with rising temperatures.

Steven Oberbauer, of Florida International University, is collaborating with the Clarks, supervising studies of the movement of carbon dioxide between La Selva and the sky above. Mounted on the tower that thrusts out of the center of the forest is high-tech equipment used in eddy flux analysis, in which measurements of wind speed and direction are combined with data on carbon dioxide concentration to track the flow of carbon in and out of the ecosystem. Using a measurement system designed by Hank Loescher, a forest ecologist at Oregon State University, the researchers have so far collected only a few years of data. What they've recorded supports the Clarks' findings. In cooler years, when the trees grow faster, they absorb more carbon out of the atmosphere. In hot years, they take up less carbon dioxide and breathe out more.

## Stirring Debate

Laboratory experiments established long ago that plants increase their rate of photosynthesis as temperatures rise. But at some critical temperature, they hit a physiological wall, and photosynthesis crashes. The idea that intense heat can sap a plant's productivity, says Clark, is "Ecophysiology 101." Yet when the Clarks and Keeling coauthored a paper on their matching results in Proceedings of the National Academy of Sciences in 2003, their analysis caused a stir.

Some researchers take exception to the Clarks' argument because it extrapolates from a small remnant of rainforest to all the forests of the tropics. Loescher points out that this forest is unique: an island of trees surrounded by farms, residences, and tourist developments. "I strongly suspect that all this land-use change makes for hotter and drier air masses that affect La Selva, contributing toward the alteration of carbon exchange and plant productivity when compared to undisturbed forests," he says. Loescher notes that the three years of eddy flux data he collected there follow the observed pattern of changing tree growth. But he believes the Clarks oversimplify when they tie the pattern only to fluctuating temperatures. In his view, drier conditions in hotter years, along with increased cloud cover in the wet season during El Niño years, contribute to the drop in plant productivity. "Just looking at temperature is fairly simple and does not have a lot of explanatory power, though the pattern is compelling."

Steven Wofsy, an atmospheric chemist at Harvard who has studied carbon exchange in the Amazon, shares some of Loescher's reservations. "The Keeling–La Selva correlation implies that La Selva is a good proxy for global tropical forests," he says. "That's plausible, but hardly proven." Wofsy's studies in the Amazon were not as long-standing as the Clarks' data set, but the pattern of change documented there did not mirror La Selva's. "When we first got there our site was losing carbon," he says, "but over the five years we were there, it turned around into a carbon sink." The trend was part of a response to past disturbance, as smaller trees shot up fast in the openings left by fallen forest giants.

Deborah Clark knows that her theory is far from ironclad. "There is the $n = 1$ problem: You should never leap off into space based on one sample," she says. "There are lots of reasons to say it would be ridiculous to extrapolate from this forest to the world tropics. Yet we've got this incredible parallel between how trees react to yearly climate variation here and what the Keeling model says the world tropics are doing. I have to think there's a big signal out there, related to rising temperatures."

The Clarks' evidence contradicts the idea that tropical forests act as a major sink for excess atmospheric carbon dioxide. Laboratory experiments had shown that increased levels of carbon dioxide could lead to impressive changes in plant growth rates: Some crops showed a 33 percent jump in productivity in response to a doubling of carbon dioxide concentrations. The Intergovernmental Panel on Climate Change (IPCC), in its 2001 report, optimistically endorsed the idea that "$CO_2$ fertilization" would result in forests and farmlands sucking more and more carbon out of the air.

In the late 1990s, researchers working in the tropics found evidence that forests were responding to increasing levels of atmospheric carbon dioxide with a burst of new growth. Oliver Phillips, of the University of Leeds, headed a group of researchers who reviewed long-term growth data from South American tropical forests. In a report published in Science in 1998, they concluded that in most of the study sites, growth exceeded losses from tree death, and that trees had packed on about one metric ton of carbon per hectare per year in recent decades, acting as a major sink for excess carbon dioxide.

Meanwhile, eddy flux studies in the Amazon, the largest intact tropical forest in the Americas, were yielding numbers that showed far more carbon dioxide was going into the ecosystem than was coming back out. Some researchers concluded that tropical forest uptake could account for all of the mysterious "missing carbon sink"—the difference between the amount of carbon dioxide pumped into the atmosphere each year by human activities and the amount that actually stays there. In some years, this sink amounts to 50 percent of total emissions.

But if Amazon forests were sucking down a ton of carbon per hectare every year, says Clark, "that forest would end up looking like the Empire State Building. Something just doesn't add up."

In November 2003, Wofsy and his colleagues published a paper in *Science* reporting the results of their

*Where forests play a major role in carbon dioxide uptake, the heavier isotope, carbon-13, builds up in the atmosphere. On the basis of shifting carbon isotope ratios, forests in the Northern Hemisphere have been estimated to absorb more than two billion metric tons of carbon every year.*

eddy flux studies at Tapajos National Forest in Brazil. It was the first project to record a net loss of carbon from a forest in the Amazon. One crucial difference between this study and previous work was in the way the scientists interpreted their eddy flux data.

Wofsy explains that valid eddy flux data depend on the constant movement of air. Mounted at the top of an eddy flux tower is a sonic anemometer, a device that measures wind speed and direction 10 times every second. These numbers are linked with carbon dioxide concentrations, measured once every second. "If the air is not moving," he says, "that experiment doesn't work very well."

At night in the deep forest, air tends to become still. In the dark, photosynthesis ceases, but plants continue to breathe out carbon dioxide, a factor that can be accounted for only if researchers correct for the lack of nighttime winds. "The first long-term eddy flux study was begun by me and my group at Harvard Forest in 1989 and is still ongoing," says Wofsy. "We published results in 1993 which showed that you need to carefully correct for time periods when the atmosphere is very stable." The researchers who'd been claiming a major carbon sink in the Amazon had failed to make this correction, and so they missed much of the carbon being exhaled by the forest.

## Forest Succession

I follow Leo Campos and William Miranda, two Costa Rican research assistants who have been working with the Clarks for more than a decade. They move deftly through the tangled forest, working on the annual census of tree growth that has been tracked for 25 years. Here on the forest floor, life for La Selva's trees is revealed as a never-ending competition for sunlight. Seedlings no thicker than my pinky finger may be 10 or 20 years old, unable to grow much in the understory until a branch or a whole tree falls, leaving a precious gap in the solid roof of leaves. Some trees that have at last reached the canopy are more than 200 years old but surprisingly slender: I could easily wrap my arms around the trunks.

To measure growth on the bigger trees, Campos climbs a ladder to measure the girth of the trunk above the thick buttresses typical of many tropical species. These radiate out from the tree's base like spokes on a massive wheel; the width of the buttresses increases much faster than the overall biomass of a tree. Deborah Clark believes some of her colleagues overestimated tree growth in their studies because they made the mistake of measuring around buttresses.

Phillips and his colleagues were trying to do "a very cool thing," says Clark. "But they were taking any data they could find. We know for a fact that many people were measuring trees in the tropics wrong, measuring around buttresses. They're very wide and grow very fast and really have zip to tell about how the tree is doing."

In 2004, the Phillips group published a second study, in which such faulty data sets had been weeded out. Once again, they concluded that tropical forests in the Americas are a carbon sink. Based on tree growth data from 59 study plots in Amazonia, they estimate that tropical forests are taking up about 0.9 metric tons of carbon per hectare each year. Still, in a review paper in *Philosophical Transactions of the Royal Society* in February 2004, Phillips and coauthor Yadvinder Malhi emphasized that a tropical carbon sink should not be taken for granted. "There is some evidence that intact tropical forests may be increasing in biomass and acting as a moderate carbon sink. . . . Almost all researchers agree that this sink cannot be relied on and may even reverse in the coming century," they wrote.

"I've never believed that tropical forests were a major carbon sink," says Wofsy. That theory ignores the constant disturbances innate to tropical forests, which he's experienced firsthand. His research in Brazil ended after repeated tree falls destroyed three expensive eddy flux towers. Tropical trees, once they're free to grow in an open patch of sun, shoot up much faster than those in temperate forests, so studies that measure only tree growth might conclude that recently disturbed, regrowing patches are a net carbon sink. But since dead wood is rotting and releasing carbon on the forest floor, tracking tree growth without also tracking carbon flux in the air gives an incomplete picture. "If we had only measured the growth of trees, and not studied net carbon exchange using eddy flux, we'd have believed our forest was a big sink for carbon, because our trees were growing like crazy—as is often true when a forest is released by having trees knocked over."

Wofsy is one of the authors of a new IPCC report on global climate change and its social and ecological impacts. "The current report does not endorse the idea of a big carbon sponge in the tropics," he says. The report does discuss some results that suggest significant carbon uptake in the tropics, averaging in the range of 0.1 to 0.2 metric tons of carbon per hectare per year. Wofsy and many of his colleagues now believe that much of the "missing carbon" is being absorbed by forests in northern latitudes—throughout North America and Eurasia, and in boreal forests that are expanding north into tundra landscapes as the climate continues to warm.

That idea is supported by remote sensing data that suggest forest biomass is expanding in the north. Another important clue comes from studies of the isotopic makeup of carbon drifting in the atmosphere. When plants take up carbon dioxide, they leave a signature in the pool of gas left behind. Plants prefer to use carbon-12, a lighter form of the atom. Where forests play a major role in carbon dioxide uptake, the heavier isotope, carbon-13, builds up in the atmosphere. On the basis of shifting carbon isotope ratios, forests in the Northern Hemisphere have been estimated to absorb more than two billion metric tons of carbon every year.

That result makes sense, says Wofsy, because virtually every temperate forest in the Northern Hemisphere has been logged or cleared for agriculture within the last two centuries. Young trees absorb carbon much faster than mature stands. But well into the process of forest succession, temperate and northern forests remain hungry for carbon.

At Harvard Forest, which is now 85 years old (the old growth was leveled by a hurricane in 1938), the rate of carbon uptake recently doubled. "That was really unexpected," says Wofsy, "but the reason became clear as we studied the changing structure of the forest Red oak is now becoming dominant over earlier successional species like red maple. The oaks have denser wood and can grow bigger than the maples." The same kind of pattern is likely to hold in the regenerating conifer forests of the West, and mature forests in the Northern Hemisphere continue to absorb carbon much longer than climate modelers previously assumed. Remnants of old-growth Douglas-fir forest in the Pacific Northwest—aged 500 years or more—continue to absorb carbon.

## Carbon Banks

At La Selva, the Clarks, Oberbauer, Loescher, and their colleagues are engaged in an ambitious project they've dubbed "Carbono," what Deborah Clark describes as the "moonshot effort" to track carbon as it flows through the intricate channels of the tropical rainforest. At each of 71 sites throughout the forest, they've erected a portable tower and used it as a platform to measure biomass in every leaf, vine, and branch from the muddy earth to the top of the canopy. But what is above ground is only the beginning.

Most of the carbon stored in tropical forests lies beneath the earth, in live roots that grow and respire, in the rich detritus of fallen leaves and wood that feed soil microbes, and locked in the soil itself. At Carbono plots scattered throughout La Selva, researchers are gathering and weighing leaf litter, photographing the growth of tree roots with underground cameras, and encapsulating bits of the forest floor to measure carbon dioxide as it flows in and out.

Six plots also include a perfectly symmetrical, three-meter-deep, hand-dug shaft. These research pits look like out-sized graves but have the pleasant aroma of a root cellar. They allow the Carbono crew to measure carbon exchange and storage deep in the ground—and there's plenty of carbon to track. "This forest holds something like 300 to 400 tons of carbon per hectare," says Clark. "If you cut it down, you'll lose carbon to the atmosphere far faster than the trees respire it, even in the hottest El Nino year." The razing and burning of tropical forests cleared for agriculture is a major contributor to rising global carbon dioxide levels. Even if forests throughout the tropics grow less and respire more in the hot years to come, they remain an irreplaceable storage bank for carbon.

# 5

# The Oceans and Weather

Every rainy season, the Guna people living on the Panamanian white sand archipelago of San Blas brace themselves for waves gushing into their tiny mud-floor huts. Once rare, flooding is now so menacing that the Guna have agreed to abandon ancestral lands for an area within their semi-autonomous territory on the east coast of the mainland.

# In Hot Water: Weather and Global Warming

The world's weather patterns and the world's oceans are linked in a dynamic equilibrium. The water that makes up the earth's oceans provides the primary driving force for global weather. However, the relationship between the oceans and weather is very complex, involving the physical behavior of water as determined by its molecular structure, the rotation and tilt of the planet, the influx of solar energy, and the hydrologic cycle.

## The Physical Behavior of Water

The water molecule is a simple combination of one oxygen atom bonded to two separate hydrogen atoms, producing a nonlinear molecular shape. The chemical structure of water gives it unusual physical properties that enable it to exist on earth in three different physical forms simultaneously: as liquid water, water ice, and water vapor. All three forms have unique characteristics that play significant roles in the relationship between the oceans and the weather.

When liquid water becomes sufficiently cold, its molecules organize into a stable crystal structure, wherein they maintain a greater separation from each other. When that happens, the volume occupied by a specific amount of liquid water increases by about 10 percent. Ice is less dense than liquid water, and so it floats. Under increasing pressure, as is exerted by the accumulation of snow and ice that form glaciers, the molecules are forced into slightly different arrangements that result in the formation of different kinds of ice. Some types of ice resemble diamond in hardness and clarity. As temperature increases and pressure decreases, ice becomes liquid water. As temperatures continue to rise, water molecules undergo a conversion from a liquid state to a gaseous state. This conversion requires water molecules to gain a large amount of extra energy relative to other materials of similar molecular structure. Pure water boils at a temperature of 100°C (212°F)—as water is heated, its temperature rises linearly. However, at 100°C, the temperature rise ceases and remains constant. The water does not automatically convert to the gaseous state; rather, a significant amount of extra energy input is required to convert 100°C liquid water into 100°C water vapor. That phenomenon is known as the "latent heat of vaporization." Though it is not as pronounced, a similar effect, called the "latent heat of fusion," exists in the conversion of liquid water at 0°C and water ice at 0°C. That effect enables water in the oceans to absorb huge amounts of solar energy and thermal energy from the earth's interior. In the atmospheric equilibrium, it has paradoxical functions. On one hand, water's ability to absorb and release energy moderates the atmosphere and prevents the extreme temperature fluctuations that would occur if the water were not present. On the other hand, energy released or absorbed by ocean waters can add fuel to storm systems over ocean waters, intensifying their natural power.

Water expands and contracts with changes in temperature. Liquid water increases in volume as its temperature rises. Scientists attribute some 40 percent of rising oceanic water levels in recent decades to the increased volume occupied by water that has become only slightly warmer than past temperature levels. Because water is a fluid, it moves according to the shape of its container and the forces exerted against it. In a tropical storm, for example, much of the storm surge is attributed to the water under the area of low atmospheric pressure being pushed upward by the higher atmospheric pressure acting on the water around it. The phenomenon is comparable to the manner in which water is forced up a drinking straw when the air pressure inside the straw is reduced. Every molecule of water in the world's oceans and connecting freshwater systems is bound to all other water molecules in the world. The movement of liquid water in one area induces a reciprocal movement of liquid water in the rest of the system.

In a gaseous state, water molecules contribute to the greenhouse effect, without which the earth would have a significantly colder normal temperature and would effectively turn it into a large ball of ice orbiting the sun. In the greenhouse effect, solar radiation that has been absorbed during daylight hours is reemitted from the earth's surface as infrared radiation. Water molecules in the atmosphere absorb some of that radiant heat and become thermally excited. When they return to their normal unexcited state, they release that energy back into the atmosphere, and in the atmospheric equilibrium system, that is sufficient to maintain the planet's above-freezing state.

In the solid state, as ice and snow, water is highly reflective. Incoming solar energy is reflected by ice and snow back out into space rather than being absorbed. That reflection effect also protects against excessive fluctuations in weather conditions.

## Planetary Rotation and Weather

The fluid nature of water makes it subject to the process of convection. Liquid water becomes less dense when it is warmed and consequently rises through cooler water. As it loses energy and becomes colder, the density of liquid water increases, and it sinks to lower depths. The planetary atmosphere also behaves in that way as air—including the water vapor contained within it—becomes less dense and rises when it absorbs heat and sinks when it loses heat and becomes denser. As the planet spins about its axis, the atmosphere does not spin consistently with it but rather is affected by various phenomena. Those include the Coriolis effect, friction with the planetary surface, the density gradient that exists throughout the air column, localized temperature differences, and a variety of other factors that influence the movement of the air. As the earth spins, air currents such as the jet stream and the trade winds are generated consistently. Though such currents have existed for millennia, they are nevertheless erratic and unpredictable in their local movement.

The long-term existence of such air currents, combined with differential absorption of solar energy, has generated oceanic currents in accordance with the exchange of energy between water and air. Most of that energy exchange occurs at the interface between the liquid ocean and the gaseous atmosphere. The Gulf Stream is perhaps the best-known oceanic current. Given that water rises as it is warmed and sinks when it is cooled in a convection cycle, the movement of relatively warm currents in the upper levels of the oceans must be countered by opposing movements of colder water at deeper levels. That reality has produced what is known as the "interoceanic conveyor belt," a continuous system of currents by which relatively warm surface waters move across the world's oceans in one direction. It is carried downward as the water cools, returns at lower depths back across the oceans, and eventually drawn upward again as it is warmed. The process is slow moving and not well understood. At the surface, the most well-known sections are the Atlantic and Pacific Gyres, the roughly circular movement of the surface waters about those bodies of water.

## Energy Exchange Drives Weather

Water has the ability to absorb great amounts of energy, effectively storing it as warmer water. The additional energy contained in warm water increases the rate of evaporation, and the effect is greatly enhanced by reductions in local air pressure, as in storm systems. The movement of air over the surface of the oceans thus acquires great amounts of moisture, which is then carried farther aloft. Typically, the energy of storm systems passing over relatively warm oceanic waters increases. The increased energy enhances the cyclonic movement of the air in the storm system and, under the right conditions, generates hurricanes and typhoons over the seas and tornado systems on land.

One of the most controversial discussions in climate change science revolves around whether global warming is driving an increase in the number and severity of such storms. Evidence is far from clear in the matter, and climate scientists are strongly polarized on the issue. Here again, the observational data relating to the occurrence of such storms is not conclusive. It is possible, for example, that increases in the number of tornadoes may have more to do with an increase over time in the number of people actively looking for them and reporting them than with any actual increase in number. Similarly, observational data for hurricanes and typhoons do not indicate causality.

## El Niño–Southern Oscillation

The El Niño–Southern Oscillation (ENSO) weather system, which includes the La Niña effect, is intimately linked to the cyclic monsoon of the Indian Ocean but was long a great unknown in the field of meteorological research. Decades of data collection, study, and technological and scientific advancement have uncovered much information about how the system works. The prevailing winds around the equator and the rotation of the planet combine to build up a large mass of warm water in the western region of the Pacific Ocean. The enhanced moisture in the air pushes over the Indian Ocean, where the warm waters drive even more moisture into the air.

Over the Indian subcontinent, the air mass, moving inland, is forced to rise and as it does so, the moisture within it condenses and falls in the form of summer monsoon rains. With the seasonal change, colder, heavier air pushes seaward from northern reaches, striking the relatively warm, humid air and triggering the winter monsoon rains. The El Niño mass remains stationary in the western Pacific as the monsoon cycle progresses, until a tipping point is reached and the mass is no longer stable. It essentially collapses and migrates eastward toward South America. The enhanced moisture content in the air above the mass moves inland, producing extra quantities of rain across much of the Western Hemisphere. The massive air movement also influences the motion of the jet stream so that weather as far away as Europe and Africa is affected. The arrival of the relatively warm water of El Niño at the coast of South America effectively shuts down the ocean currents that function there, greatly affecting the lives of the people who live in the region.

El Niño's twin, La Niña, represents its systematic opposite. Warm waters move eastward across the Pacific once again, drawing up cold deep ocean waters along the South American coast. The colder, drier air over those cold-water currents does not produce enhanced rainfall. In fact, opposite effects are observed globally, as reduced rainfall and drought conditions are triggered throughout the Western Hemisphere and as far away as eastern Africa and southern Europe.

Until the magnitude of the ENSO system was realized, the weather effects that it drives throughout the world were thought to be entirely localized events. One of the great debates in climate change is the effect of global warming on the system and the long-term ramifications of such effects.

# After Sandy: Why We Can't Keep Rebuilding on the Water's Edge

By Bryan Walsh
*Time,* November 20, 2012

It's so obvious we forget it: an extreme-weather event becomes a disaster only if it hits where people and their possessions are. Of the 19 tropical storms that were tracked during this summer's Atlantic hurricane season, 10 veered off harmlessly into the Atlantic Ocean, never making landfall. But when a storm like Sandy tracks over the most heavily populated stretch of land in the western hemisphere, the damage to people and property can be immense. Sandy wasn't the strongest storm—it was just barely a Category 1 hurricane when it made landfall at the end of last month—but both its death toll and its economic damage were high simply because so many people were in its path. Storm plus people equals natural disaster. The hurricane is the spark, but population is the tinder.

That's why, as the Northeast begins the long process of rebuilding, we need to think about what we can do to minimize the number of people and the value of the property that might be in the way of the next storm. So far, most of that discussion has settled around the possibility of building multibillion-dollar seawalls and barriers that might be able to shield Manhattan and other vulnerable places from the kind of storm surges that caused so much destruction during Sandy. Seawalls do have their place—the Connecticut town of Stamford escaped major damage thanks in part to its own barrier—especially as the climate warms and seas rise. But if people didn't live in so many high-risk places, we wouldn't have to put any protective infrastructure there at all.

The reason so many Americans make their homes in storm and flood zones is partly because we simply like living along the water. But the other part is that government-subsidized flood insurance essentially eliminates the financial risk. The question now, after Sandy, is whether we'll keep making the same circular mistake, paying billions to put people back in harm's way, or whether we'll instead say, "Build if you want, but the risk is all yours."

The Northeast spots that were most heavily damaged by Sandy—and, sadly, the areas where the most lives were lost—were in housing developments that were built very close to the coast, places like Staten Island and Breezy Point in New York, and Ocean County on the New Jersey shore. As the *Huffington Post* made clear in a deeply reported piece last week, those same areas had seen dramatic development over the past couple of decades, despite the fact that government officials knew that the coastal land would be vulnerable to flooding from a major storm:

> Given the size and power of the storm, much of the damage from the surge was inevitable. But perhaps not all. Some of the damage along low-lying coastal areas was the result of years of poor land-use decisions and the more immediate neglect of emergency preparations as Sandy gathered force, according to experts and a review of government data and independent studies.
>
> Authorities in New York and New Jersey simply allowed heavy development of at-risk coastal areas to continue largely unabated in recent decades, even as the potential for a massive storm surge in the region became increasingly clear.
>
> In the end, a pell-mell, decades-long rush to throw up housing and businesses along fragile and vulnerable coastlines trumped commonsense concerns about the wisdom of placing hundreds of thousands of closely huddled people in the path of potential cataclysms.

States like New York and New Jersey were hardly alone in packing people along the coastlines: a 2005 report from Princeton University noted that nearly a quarter of the world's population lives less than 328 ft. (100 m) above sea level and within 60 miles (97 km) of the coast. But *Huffington Post* notes that Ocean County's population rose nearly 70 percent from 1980 to 2010, and in that final year alone, more residential building permits were issued in that county than in any other in New Jersey. In New York as well, hundreds of new structures have gone up in the high-risk storm-surge area of Staten Island.

Human beings have been crowding along the coasts for as long as they've been building cities, and air travel and the Internet haven't made ports obsolete yet. But in the past, those who made the choice to live near the ocean also knew to treat its immense power with respect, even building their homes facing the land. Today we're much more reckless, and an ocean view is worth paying extra for.

As Justin Gillis and Felicity Barringer wrote in the *New York Times* this week, the federal government is bound not just by elective policy but also by law to pay for most of the cost of fixing storm-damaged infrastructure—including homes. Add in the National Flood Insurance Program, which offers consumers in coastal danger zones below-market protection from floods, and you can see how the federal government is almost making it easier to live in a danger zone than to make the hard choice of relocating:

> **The question now, after Sandy, is whether we'll keep making the same circular mistake, paying billions to put people back in harm's way, or whether we'll instead say, "Build if you want, but the risk is all yours."**

Across the nation, tens of billions of tax dollars have been spent on subsidizing coastal reconstruction in the aftermath of storms, usually with little consideration of whether it actually makes sense to keep rebuilding in disaster-prone areas. If history is any guide, a large fraction of the federal money allotted to New York, New Jersey and other states recovering from Hurricane Sandy—an amount that could exceed $30 billion—will be used the same way.

Tax money will go toward putting things back as they were, essentially duplicating the vulnerability that existed before the hurricane.

The federal government's flood-insurance program is already under major financial strain, and Sandy could cost as much as $7 billion just in terms of government insurance claims, while the program itself is only allowed to add an additional $3 billion to its currently high debt levels. That might prompt lawmakers to finally reform the program—and if subsidized flood insurance is no longer a given, we might also begin to see a slowdown in coastal population growth. That's long overdue. We can try to reduce climate change and we can try to build physical protections for established coastal population centers. But the best way to ensure that the next Sandy does less damage is simply to keep people out of harm's way—or at least make it more expensive to stay there.

# How Climate Change Threatens the Seas

By Dan Vergano
*USA Today,* March 13, 2012

The tide rolls out on a chilly March evening, and the oystermen roll in, steel rakes in hand, hip boots crunching on the gravel beneath a starry, velvet sky.

As they prepare to harvest some of the sweetest shellfish on the planet, a danger lurks beyond the shore that will eventually threaten clams, mussels, everything with a shell or that eats something with a shell. The entire food chain could be affected. That means fish, fishermen and, perhaps, you.

"Ocean acidification," the shifting of the ocean's water toward the acidic side of its chemical balance, has been driven by climate change and has brought increasingly corrosive seawater to the surface along the West Coast and the inlets of Puget Sound, a center of the $111 million shellfish industry in the Pacific Northwest.

*USA Today* traveled to the tendrils of Oyster Bay as the second stop in a year-long series to explore places where climate change is already affecting lives.

The acidification taking place here guarantees the same for the rest of the world's oceans in the years ahead. This isn't the kind of acid that burns holes in chemist's shirt sleeves; ocean water is actually slightly alkaline. But since the start of the industrial revolution, the world's oceans have grown nearly 30 percent more acidic, according to a 2009 Scientific Committee on Oceanic Resources report. Why? Climate change, where heat-trapping carbon dioxide emitted into the air by burning coal, oil and other fossil fuels ends up as excess carbonic acid absorbed into the ocean.

That shift hurts creatures like oysters that build shells or fish that eat those creatures or folks like shellfish farmer Bill Dewey, who makes his living off the ocean.

"As fresh as they get, you could eat one now," says Dewey of Taylor Shellfish Farms in Shelton, Wash., shucking an oyster open, mud running from its shell to reveal the opulent meat within, silver and white in the starlight. The black lip curling around the sweet-tasting shellfish reveals it to be a Pacific oyster, farmed worldwide.

"Folks think we just get rich picking oysters off the ground. A lot of work goes into every one of these, and we can't afford to lose any of them," Dewey says.

Lose them they have, and lose them they will, to the water lapping at Dewey's hip boots where the low tide meets the flats.

"We are looking into the future happening now," Dewey says. And researchers are seeing similar corrosive effects on Florida's coral reefs that shelter young fish and on the tiny sea snails that feed salmon and other species in the Pacific Ocean.

## Too Much of "a Good Thing"

Because of ocean chemistry, water three times more acidic resides at greater ocean depths. When conditions are right, strong winds blowing over ocean water along steep coasts, such as along the West Coast of North America, generate "upwelling" of these deep waters. The results bring this more corrosive seawater to the shallows of places such as Puget Sound, a foreshadowing today of how the oceans will look in a few decades.

"We are able to see the effects of ocean acidification," says oceanographer Richard Feely of the National Oceanic and Atmospheric Administration. He first charted upwelling of deeper, corrosive ocean water on the surface of the Pacific Ocean along the West Coast on a 2007 expedition.

How did it get there in the first place? Atmospheric carbon ends up absorbed directly by the ocean, where plankton sucks up carbon dioxide via photosynthesis. "That's a good thing," Feely says, because the carbon dioxide they ingest means less warming of the atmosphere.

But when those sea plants and creatures die, they fall to the depths. Some of the consumed carbon ends up dissolved in deep ocean waters.

"Once we had the canary in the coal mine; now we have the oyster in the ocean," Washington Gov. Jay Inslee says.

## An Oyster Mystery

In the same year that Feely and his colleagues made their upwelling discovery, disaster struck the Whiskey Creek Shellfish Hatchery in Netarts Bay, perched on the edge of Oregon's Pacific Coast. Baby oysters grown there to be shipped to shellfish farms worldwide were dying en masse, tens of millions of millimeter-size larvae filling hatching tanks, all dead.

"We were very close to going out of business. It was a major deal," says Sue Cudd of the Whiskey Creek hatchery. Bankruptcy threatened because of unfilled orders, and the entire industry faced collapse without the larvae, the seeds that shellfish growers needed to raise Pacific oysters over the next 18 months. A second hatchery run by Dewey's employers, Taylor Shellfish—one of four in the region—saw the same die-off the next year, with the same failure to spawn striking wild oysters in Washington's Willapa Bay, a longtime source of seeds for the industry.

It was only at the end of 2008 at a shellfish farmer's meeting that Feely delivered the bad news: On days in the summer along the West Coast when northerly winds blew, deep ocean water was likely dissolving the fragile young oysters.

Without industrial emissions of greenhouse gases, the upwelling wouldn't be a problem because really corrosive water wouldn't get high enough to reach the surface, says Burke Hales of Oregon State University in Corvallis. "That extra kick from (man-made) carbon is what pushes the saturation point high enough," Hales says. "Anyone who says this is a natural thing just doesn't get it or is missing that key point."

Shellfish rely on calcium to build their shells, and a more corrosive water makes that harder for them to the point of becoming impossible.

Alan Barton, the staff scientist at Whiskey Creek, realized the seawater was far too acidic and started making phone calls to Hales to have the water examined. Last year, Barton, Hales and colleagues, including Feely, definitively showed that those deep waters do dissolve baby oysters only a few days old, ones which rely on a very soft form of calcium for their initial growth spurt.

"A lot of people out here I talk to don't believe in climate change, but ocean acidification—to them that's real—because they can see it eating into their livelihoods," Hales says. "The chemistry is really simple and really inevitable: More carbon in the air means more carbon in the ocean, and there is no getting around it."

## An Ocean-Size Challenge

Benoit Eudeline, staff scientist for Taylor Shellfish, and colleagues such as Barton at Whiskey Creek have become evangelists on the topic for the seafood industry, warning their colleagues on the East Coast of what is coming. There, deep-water upwelling is not a problem, but warmer waters have shifted crab and fish populations while acidic mud on the Maine seafloor has hurt clamming.

> *"That extra kick from (man-made) carbon is what pushes the saturation point high enough,"* Hales says. *"Anyone who says this is a natural thing just doesn't get it or is missing that key point."*

"We were joking about what size antacid tablet you would need to fix the ocean," Eudeline said. "It is so tremendous that you can only make jokes about it."

That's because the ocean absorbs 23 percent of all man-made carbon dioxide emissions, according to a 2012 *Earth System Science Data* journal report, more than 8 billion tons of the stuff every year. It would take a very large Alka-Seltzer tablet to fix that. "That's why talk about just avoiding warming from here on out in the atmosphere is missing the effects that are already locked in due to ocean acidification," Feely says.

Back on the tidal inlet at Oyster Bay, Dewey walks along rotting wooden dike walls placed there more than a century ago, by earlier shellfish farmers, down to where the low tide has fallen. All kinds of shellfish and other sea life, including the famed salmon of the Pacific Northwest, live in these same waters, enjoying the bounty of these waves.

"We only know what is happening to shellfish because they have spokesmen," Dewey says. "Only the good Lord knows what is happening to everything else out there."

# Amount of Coldest Antarctic Water Near Ocean Floor Decreasing for Decades

### NOAA, March 20, 2012

Scientists have found a large reduction in the amount of the coldest deep ocean water, called Antarctic Bottom Water, all around the Southern Ocean using data collected from 1980 to 2011. These findings, in a study now online, will likely stimulate new research on the causes of this change.

Two oceanographers from NOAA and the University of Washington find that Antarctic Bottom Water has been disappearing at an average rate of about eight million metric tons per second over the past few decades, equivalent to about fifty times the average flow of the Mississippi River or about a quarter of the flow of the Gulf Stream in the Florida Straits.

"Because of its high density, Antarctic Bottom Water fills most of the deep ocean basins around the world, but we found that the amount of this water has been decreasing at a surprisingly fast rate over the last few decades," said lead author Sarah Purkey, graduate student at the School of Oceanography at the University of Washington in Seattle, Wash. "In every oceanographic survey repeated around the Southern Ocean since about the 1980s, Antarctic Bottom Water has been shrinking at a similar mean rate, giving us confidence that this surprisingly large contraction is robust."

Antarctic Bottom Water is formed in a few distinct locations around Antarctica, where seawater is cooled by the overlying air and made saltier by ice formation. The dense water then sinks to the sea floor and spreads northward, filling most of the deep ocean around the world as it slowly mixes with warmer waters above it.

The world's deep ocean currents play a critical role in transporting heat and carbon around the planet, thus regulating our climate.

While previous studies have shown that the bottom water has been warming and freshening over the past few decades, these new results suggest that significantly less of this bottom water has been formed during that time than in previous decades.

*Changes in the temperature, salinity, dissolved oxygen, and dissolved carbon dioxide of this prominent water mass have important ramifications for Earth's climate, including contributions to sea level rise and the rate of Earth's heat uptake.*

"We are not sure if the rate of bottom water reduction we have found is part of a long-term trend or a cycle," said co-author Gregory C. Johnson, Ph.D., an oceanographer at NOAA's Pacific Marine Environmental Laboratory in Seattle. "We need to continue to measure the full depth of the oceans, including these deep ocean waters, to assess the role and significance that these reported changes and others like them play in the Earth's climate."

Changes in the temperature, salinity, dissolved oxygen, and dissolved carbon dioxide of this prominent water mass have important ramifications for Earth's climate, including contributions to sea level rise and the rate of Earth's heat uptake.

"People often focus on fluctuations of currents in the North Atlantic Ocean as an indicator of climate change, but the Southern Ocean has undergone some very large changes over the past few decades and also plays a large role in shaping our climate," said Johnson.

The data used in this study are highly accurate temperature data repeated at roughly 10-year intervals by an international program of repeated ship-based oceanographic surveys (www.goship.org). Within the US, the collection of these data has been a collaborative effort of governmental laboratory and university scientists, funded primarily by NOAA and the National Science Foundation. However, much of the data used in this study were measured by international colleagues.

"Collection of these data involves 12-hour days, seven days a week, of painstaking, repetitive work at sea, often for weeks on end with no sight of land. We are grateful for the hard work of all those who helped in this effort," said Purkey.

# Blue Crabs in Maine? Something Fishy About Global Warming

By Pete Spotts
*Christian Science Monitor,* May 16, 2013

*Warming oceans are changing the mix of species in the world's fisheries as fish try to remain in waters in their preferred temperature range, according to a new study.*

The movement to keep pace with preferred temperatures shows up most starkly in the northeastern Pacific Ocean and the northeastern Atlantic Ocean, as fish migrate out of the subtropics to beat the heat.

The changes have particular implication for people living in the coastal tropics who either subsist on fishing or fish commercially, the research team says. If ocean temperatures continue to warm there, the heat could top a level that even tropical species find intolerable, reducing their abundance, the researchers say.

This raises the urgency of adopting approaches that minimize other stresses on fisheries, such as pollution and overfishing, the team says.

Marine-ecosystem models have indicated that global warming's impact on ocean temperatures would trigger such a migration. And studies of individual regions have documented the arrival of species from warmer aquatic climes.

This latest effort represents the first attempt at documenting the changes for the planet as a whole, says William Cheung, a scientist with the Fisheries Centre at the University of British Columbia in Vancouver, who led the team. The techniques that the team used, along with the results, appear in Thursday's issue of the journal *Nature*.

The general pattern reported in the study is "very similar" to results from studies that have focused on the US Northeast's fisheries, says Michael Fogarty, who heads the ecosystem assessment program at the National Oceanic and Atmospheric Administration's Northeast Fisheries Science Center in Woods Hole., Mass.

Off the New England coast, for instance, marine scientists tracked

> **The team then compared changes in catch temperature with changes in sea-surface temperature. When the researchers did that, they found a strong correlation between rising catch temperatures and rising sea-surface temperatures in the same region.**

migration trends for 36 fish species and found that 75 percent had moved north or into deeper water or both to keep their cool, Dr. Fogarty says.

At the same time, "the Atlantic croaker, a subtropical species, is moving north and is having higher reproductive success as well" in northern waters, he says.

Meanwhile, fishermen in the Gulf of Maine are reporting highly unusual species for the area: black sea bass, which could earn them a tidy sum; new species of squid; and blue crabs, Fogarty adds.

The work by Dr. Cheung and colleagues "is a very interesting study, and its global reach makes it quite important," he says.

The study covers a period spanning 1970 to 2006. The team examined catch records compiled by the United Nations Food and Agriculture Organization, as well as from regional and national fisheries groups. The researchers divvied the catch data among 52 large marine ecosystems—for example, the US Northeast's continental shelf, the North Sea, or ecosystems defined by currents such as the Canary Current (a segment of a much larger North Atlantic surface current that skirts the Canary Islands).

The researchers used the catch information, which collectively covered 990 fish species, to determine the relative abundance of species in each of these regions for each year the study covered. They also determined each species' preferred temperature range.

From these data, they calculated a "catch temperature"—the average of the preferred temperature ranges of all the species in each of the large marine ecosystems they identified. Changes in the average catch temperature became a stand-in for changes in the mix of species.

The team then compared changes in catch temperature with changes in sea-surface temperature. When the researchers did that, they found a strong correlation between rising catch temperatures and rising sea-surface temperatures in the same region.

Between 1970 and 2006, the global-average catch temperature increased by 0.19 degrees C per decade. As global averages go, the figure masks significant regional differences. For the northeastern Pacific and the northeastern Atlantic, the catch temperature increased by a whisker under 0.5 degrees C per decade. The increases coincided with sea-surface temperatures that were increasing by 0.2 degrees C per decade in the northeastern Pacific and 0.26 degrees C per decade in the northeastern Atlantic.

Ocean temperatures in the 14 ecosystems in the tropics increased at a pace of about 0.14 degrees per decade. The catch temperature in the tropics, the researchers note, rose by 0.6 degrees between 1970 and 1980 to 26 degrees C, then leveled off. This suggested to the team that things had gotten too hot for the subtropical species that once shared these waters with the tropical fish. The subtropical species voted with their fins and headed for cooler aquatic climes.

The use of catch data for studies like this has its limits, Fogarty notes. The quality of catch records can vary widely in different parts of the world. Apart from the rigor people bring to recording their catches, catch records can change just because of different levels of effort people exert to catch fish, he says.

The researchers appear to have been aware of these and other shortcomings of catch data, he adds. But for the kind of global question the team was asking, catch data represent the most comprehensive source of information available.

Indeed, the approach could lend itself to regular updates—tracking changes in fisheries as they happen, in much the same way systematic temperature records track temperature trends, some researchers say.

The evidence of climate change's global impact on fisheries is "startling," notes Mark Payne, a researcher with the National Institute of Aquatic Resources at the Technical University of Denmark in Lyngby, outside Copenhagen.

The changes present adaptation challenges for local fishing interests in the developed and developing world, he notes in an e-mail.

For some regions, especially in the developed world, the changes may not portend effects as dramatic as California's loss of "Cannery Row" in the 1950s or the collapse of the Northeast's cod fisheries in the 1990s, which shuttered fishing villages in Newfoundland. But fishermen will have to retrace a learning curve to get to know the habits of new species that move into a fishery.

Pursuit of the traditional species into new waters could set the stage for international disputes over fishing.

The largest adjustment may be required in the tropics, the research team posits, because as subtropical fish leave for cooler waters, they aren't being replaced by fish seeking relief from still-warmer water elsewhere.

Adaptation measures could include adding other sources of income "or changing their fishing practices," Cheung says.

# Oceans Under Surveillance

By Quirin Schiermeier
*Nature,* May 1, 2013

A "global conveyor belt"stirs the oceans from top to bottom, with surface currents transporting warm water to the poles while cold water in the depths flows back to the tropics. But it operates in fits and starts, with the strength of the currents varying widely. Eager for a better understanding of how the vagaries of the conveyor belt shape weather and climate, oceanographers are planning two new large-scale projects to watch over Atlantic currents.

An array of instruments between Florida and the Canary Islands has been continuously monitoring the strength of the North Atlantic portion of the global conveyor belt since 2004. In December, if all goes well, an international project led by the United States will begin another set of continuous measurements of the Atlantic Meridional Overturning Circulation (AMOC), using an array of sensors strung between South Africa and Argentina. And this month, US and British funding agencies are set to decide whether they will support a new surface-to-bottom monitoring array between Labrador in Canada and Scotland, UK. The United Kingdom will also decide whether to continue operating the existing array.

Expanding such monitoring is crucial if scientists are to improve seasonal weather and climate forecasts, says Harry Bryden, an oceanographer at the University of Southampton, UK. Components of the AMOC, such as the Gulf Stream, ferry vast amounts of heat from the tropics to high latitudes, heating the winds that keep Europe's climate mild. As a result, year-to-year and longer-term changes in the strength of these currents can affect seasonal conditions across much of Europe, Africa, South America, and North America.

Observations from the UK-funded Rapid Climate Change monitoring array (RAPID)—the existing line of instrument-equipped moorings that measure current speed and direction, water temperature, salinity and pressure at various depths along the latitude line at 26.5° north—suggest that the strength of the overturning circulation can vary enormously.[1] In April 2009, the array recorded[2] a 30 percent drop in average current strength that persisted for a year, reducing the amount of heat transported to the North Atlantic by almost 200 trillion watts—equal to the output of more than 100,000 large power plants.

The anomaly—much bigger than any change that models suggested could happen—was driven by unusual wind patterns, strengthening of warm surface currents and weakening of cold water flows in the deep ocean. It has been linked to the unusually harsh winter in Europe in 2009–10. Bryden wonders whether the anomaly also helped to produce unusually wet weather in the United Kingdom. "We had six lousy summers in a row in Britain," he says. "What's going on?"

To investigate, scientists are now focusing on a crucial component of the conveyor belt: the region of the North Atlantic in which surface water heading north from the tropics cools and sinks before it moves back towards the equator. Climate models suggest that the rate of this formation of deep water will decrease by the end of the century.[3] That is problematic not only because deep-water formation drives the ocean circulation, but also because it carries vast amounts of carbon dioxide to the depths, sequestering it from the atmosphere.

"We need to find out how water masses at high latitudes are tied to the larger Atlantic circulation," says Susan Lozier, a physical oceanographer at Duke University in Durham, North Carolina. "That is not only of interest to oceanographers. The ocean moves such huge amounts of heat and carbon around that most everyone should care."

To understand how deep-water formation works, and why it varies, Lozier and her colleagues have proposed setting up an array of moored instruments and autonomous gliders called the Overturning in the Subpolar North Atlantic Program (OSNAP). This consists of two legs: a western line extending from southern Labrador to the southwest tip of Greenland, and an eastern line from Greenland to Scotland. If the US National Science Foundation and the UK Natural Environment Research Council approve the US$24-million project, measurements of heat and currents in the deep-water-formation region could start in July 2014. They are expected to give their decision later this month. If the array goes ahead, Canada, Germany and the Netherlands have all promised to contribute instruments to it.

Scientists are also trying to trace the cold, deep water as it flows into the turbulent South Atlantic, which also receives an influx of warm surface water from the Indian Ocean. South Africa, Brazil, France, Argentina and the United States are all contributing to a monitoring array that is being built at 34.5° south, between South Africa and Argentina, as part of the US$5-million South Atlantic Meridional Overturning Circulation (SAMOC) program.

By the end of the year, if all goes well, a network of bottom-moored instruments will begin to record water temperature and salinity at different levels in the deep, cold currents that run along the edges of the ocean basin. By combining those data with acoustic measurements of current velocity and bottom pressure, and with temperature and salinity data recorded by freely drifting profiling floats in the open ocean, scientists should be able to calculate the strength of the overturning circulation at that latitude,

> *To investigate, scientists are now focusing on a crucial component of the conveyor belt: the region of the North Atlantic in which surface water heading north from the tropics cools and sinks before it moves back towards the equator.*

says Silvia Garzoli, chief scientist at the Atlantic Oceanographic and Meteorological Laboratory in Miami, Florida, and a member of the project's executive committee.

Most scientists regard the idea that global warming will trigger a collapse of ocean circulation—the apocalyptic scenario that inspired the 2004 action film *The Day After Tomorrow*—to be exceedingly unlikely. But Bryden says that the 2009 Atlantic circulation glitch is an indication of just how surprising ocean behavior can be. "The next one," he says, "may be twice as big."

## References

1. Cunningham, S. A., *et al. Science* 317, 935–938 (2007).
2. McCarthy, G., *et al. Geophys. Res. Lett.* 39, L19609 (2012).
3. Meehl, G. A., *et al.* in *Climate Change 2007: The Physical Science Basis* (eds. Solomon, S., *et al.*) Ch. 10 (Cambridge Univ. Press, 2007).

# Encroaching Sea Already
# a Threat in Caribbean

By David McFadden
Associated Press, May 7, 2013

The old coastal road in this fishing village at the eastern edge of Grenada sits under a couple of feet of murky saltwater, which regularly surges past a hastily-erected breakwater of truck tires and bundles of driftwood intended to hold back the Atlantic Ocean.

For Desmond Augustin and other fishermen living along the shorelines of the southern Caribbean island, there's nothing theoretical about the threat of rising sea levels.

In fact, a 2007 report by the Nobel Prize–winning Intergovernmental Panel on Climate Change said the devastation wreaked on Grenada by 2004's Hurricane Ivan "is a powerful illustration of the reality of small-island vulnerability." The hurricane killed 28 people, caused damage twice the nation's gross domestic product, damaged 90 percent of the housing stock and hotel rooms and shrank an economy that had been growing nearly 6 percent a year, according to the climate scientists' report.

Storms and beach erosion have long shaped the geography of coastal environments, but rising sea levels and surge from more intense storms are expected to dramatically transform shorelines in coming decades, bringing enormous economic and social costs, experts say. The tourism-dependent Caribbean is thought to be one of the globe's most vulnerable regions.

"It's a massive threat to the economies of these islands," said Owen Day, a marine biologist with the Caribsave Partnership, a nonprofit group based in Barbados that is spearheading adaptation efforts. "I would say the region's coastal areas will be very severely impacted in the next 50 to 100 years."

Scientists and computer models estimate that global sea levels could rise by at least 1 meter (nearly 3.3 feet) by 2100, as warmer water expands and ice sheets melt in Greenland and Antarctica. Global sea levels have risen an average of 3 centimeters (1.18 inches) a decade since 1993, according to many climate scientists, although the effect can be amplified in different areas by topography and other factors.

In the 15 nations that make up the Caribbean Community bloc, that could mean the displacement of 110,000 people and the loss of some 150 multimillion-dollar tourist resorts, according to a modeling analysis prepared by Caribsave for the United Nations Development Program and other organizations. Twenty-one of 64 regional airports could be inundated. About 5 percent of land area in the Bahamas and 2 percent of Antigua and Barbuda could be lost. Factoring in surge from more intense storms means a greater percentage of the regional population and infrastructure will be at risk.

In eastern Grenada, people living in degraded coastal areas once protected by mangrove thickets say greater tidal fluctuations have produced unusually high tides that send seawater rushing up rivers. Farmers complain that crops are getting damaged by the intrusion of the salty water.

Adrian George is one of the coastal residents preparing to move into an inland apartment complex built by the Chinese government following the devastation left by Hurricane Ivan.

"I'm now ready to move up to the hills," George said in the trash-strewn eastern Grenadian village of Soubise, which is regularly swamped with seawater and debris at high tide. "Here, the waves will just keep getting closer and closer until we get swept away."

One response in the wealthier island of Barbados has been building a kilometer-long breakwater and waterfront promenade to help protect fragile coastlines. In most cases, international money is pouring in to kick-start "soft engineering" efforts restoring natural buffers such as mangroves, grasses and deep-rooted trees such as sea grape. Some call that the most effective and cheapest way to minimize the impact of rising seas.

But in the long run, "we need to move our centers of population, infrastructure, et cetera, out of the areas likely to become vulnerable to rising seas," said Anthony Clayton, a climate change expert and the director of a sustainability institute at Jamaica's campus of the University of the West Indies.

Where to rebuild will be yet another challenge, with the region's islands mostly rugged and mountainous with small areas of flat land in coastal areas.

Even with the Caribbean so threatened, many islands have been slow to adapt, and awareness of the problem has only recently grown. Last year, the European Investment Bank announced it would give $65 million in concessionary loans to help 18 Caribbean nations adapt, while conservation groups try, among other projects, to restore buffering mangroves and set up fishing sanctuaries to help fringing reefs recover. The Caribbean Community Climate Change Center in Belize is managing the regional response.

Yet not everyone is convinced that climate change is as dire as forecast.

Peter De Savary, a British entrepreneur and major property developer on Grenada's famed Grande Anse Beach, said the availability of capital, energy costs and the health of the global economy are far more imminent concerns than rising sea levels. He notes that most existing beach resorts will have to be rebuilt anyway in coming decades due to normal wear and tear so projected climate change impacts won't require much attention.

"If the sea level rises a foot or two it really doesn't make any difference here in Grenada because we have beaches that have a reasonably aggressive falloff," De Savary said. "If the water gets a few degrees warmer, well, that's what people come to the Caribbean for, warm water, so that's not an issue."

Shyn Nokta, who heads Guyana's office of climate change, said there's ample evidence the impacts will be less benign. Warming ocean waters have helped to significantly degrade the region's protective reefs, and threats to Caribbean coral are only expected to intensify as a result of ocean acidification due to greenhouse gases. Rainfall also has become increasingly erratic.

> *But in the long run, "we need to move our centers of population, infrastructure, et cetera, out of the areas likely to become vulnerable to rising seas," said Anthony Clayton.*

Many are also girding for climate change's impacts on an already fragile agriculture sector and drinking water quality and availability.

"The weather and climate system in the region is changing," Nokta said from Guyana's capital of Georgetown, which sits below sea level behind a complicated system of dikes and is extremely vulnerable to flooding.

Inequalities in income will play a big role in determining how the suffering is meted out island to island, said Ramon Bueno, a Massachusetts-based analyst who has researched and modeled climate change economic impacts for years.

"A low-income family living by the shoreline, with limited access to clean fresh water and earning a living from tourism, fishing or agriculture is vulnerable in a way that a middle- or high-income professional living in good air-conditioned housing at higher elevation inland is not," Bueno said.

That portends a dire future for people such as Allison Charles, a subsistence farmer in Grenada's coastal village of Telescope, a fact she said she's well aware of.

"It's hard now. Already our plants are getting burned by the salt water coming up the river," Charles said in her village, framed by Grenada's rugged hills. "I can't really imagine what the future will hold."

# The Coming Arctic Boom

By Scott Borgerson
*Foreign Affairs,* July/August 2013

## As the Ice Melts, the Region Heats Up

The ice was never supposed to melt this quickly. Although climate scientists have known for some time that global warming was shrinking the percentage of the Arctic Ocean that was frozen over, few predicted so fast a thaw. In 2007, the Intergovernmental Panel on Climate Change estimated that Arctic summers would become ice free beginning in 2070. Yet more recent satellite observations have moved that date to somewhere around 2035, and even more sophisticated simulations in 2012 moved the date up to 2020. Sure enough, by the end of last summer, the portion of the Arctic Ocean covered by ice had been reduced to its smallest size since record keeping began in 1979, shrinking by 350,000 square miles (an area equal to the size of Venezuela) since the previous summer. All told, in just the past three decades, Arctic sea ice has lost half its area and three quarters of its volume.

It's not just the ocean that is warming. In 2012, Greenland logged its hottest summer in 170 years, and its ice sheet experienced more than four times as much surface melting as it had during an average year over the previous three decades. That same year, eight of the ten permafrost-monitoring sites in northern Alaska registered their highest-ever temperatures, and the remaining two tied record highs. Hockey arenas in northern Canada have even begun installing refrigeration systems to keep their rinks from melting.

Not surprisingly, these changes are throwing the region's fragile ecosystems into chaos. While tens of thousands of walruses, robbed of their ice floes, are coming ashore in northwest Alaska, subarctic flora and fauna are migrating northward. Frozen tundras are starting to revert to the swamplands they were 50 million years ago, and storms churning newly open waters are eroding shores and sending the homes of indigenous populations tumbling into the sea.

No matter what one thinks should be done about global warming, the fact is, it's happening. And it's not all bad. In the Arctic, it is turning what has traditionally been an impassible body of water ringed by remote wilderness into something dramatically different: an emerging epicenter of industry and trade akin to the Mediterranean Sea. The region's melting ice and thawing frontier are yielding access to troves of natural resources, including nearly a quarter of the world's estimated undiscovered oil and gas and massive deposits of valuable minerals. Since summertime Arctic sea routes save thousands of miles during a journey between the Pacific Ocean and the Atlantic Ocean, the Arctic also stands to become a central passageway for global maritime transportation, just as it already is for aviation.

Part of the reason the Arctic holds so much promise has to do with the governments surrounding it. Most have relatively healthy fiscal balance sheets and, with the exception of Russia, predictable laws that make it easy to do business and democratic values that promote peaceful relations. The Arctic countries have also begun making remarkably concerted efforts to cooperate, rather than fight, as the region opens up, settling old boundary disputes peacefully and letting international law guide their behavior. Thanks to good governance and good geography, such cities as Anchorage and Reykjavik could someday become major shipping centers and financial capitals—the high-latitude equivalents of Singapore and Dubai.

Of course, while Arctic warming is a fait accompli, it should not be taken as a license to recklessly plunder a sensitive environment. If developed responsibly, however, the Arctic's bounty could be of enormous benefit to the region's inhabitants and to the economies that surround it. That's why all the Arctic countries need to continue their cooperation and get to work establishing a shared vision of sustainable development, and why the United States in particular needs to start treating the region as an economic and foreign policy priority, as China is. Like it or not, the Arctic is open for business, and governments and investors have every reason to get in on the ground floor.

## Much Ado about Nothing

Just a half decade ago, the scramble for the Arctic looked as if it would play out quite differently. In 2007, Russia planted its flag on the North Pole's sea floor, and in the years that followed, other states also jockeyed for position, ramping up their naval patrols and staking out ambitious sovereignty claims. Many observers—including me—predicted that without some sort of comprehensive set of regulations, the race for resources would inevitably end in conflict. "The Arctic powers are fast approaching diplomatic gridlock," I wrote in these pages in 2008, "and that could eventually lead to armed brinkmanship."

But a funny thing happened on the way to Arctic anarchy. Rather than harden positions, the possibility of increased tensions has spurred the countries concerned to work out their differences peacefully. A shared interest in profit has trumped the instinct to compete over territory. Proving the pessimists wrong, the Arctic countries have given up on saber rattling and engaged in various impressive feats of cooperation. States have used the 1982 UN Convention on the Law of the Sea (UNCLOS)—even though the United States never ratified it—as a legal basis for settling maritime boundary disputes and enacting safety standards for commercial shipping. And in 2008, the five states with Arctic coasts—Canada, Denmark, Norway, Russia, and the United States—issued the Ilulissat Declaration, in which they promised to settle their overlapping claims in an orderly manner and expressed their support for UNCLOS and the Arctic Council, the two international institutions most relevant to the region.

The Arctic powers have kept that promise. In 2010, Russia and Norway settled their long-running maritime boundary disagreement near the Svalbard Islands, and Canada and Denmark are now exploring a proposal to split Hans Island, an uninhabited rock they disputed for decades. In 2011, the Arctic countries signed a search-and-rescue agreement brokered under the auspices of the Arctic Council; this past April, they began working on an agreement to regulate commercial fishing; and this summer, they are finalizing plans for jointly responding to oil spills. Some Arctic countries are even sharing one another's icebreakers to map the seabed as part of a process, established under UNCLOS, to demarcate their extended continental shelves. Although some sticking points remain—Ottawa and Washington, for instance, have yet to agree on whether the Northwest Passage constitutes a series of international straits or Canadian internal waters and where exactly their maritime boundary in the Beaufort Sea lies—the thorniest differences have been settled, and most that remain involve areas far offshore and concern the least economically relevant parts of the Arctic.

None of this cooperation required a single new overarching legal framework. Instead, states have created a patchwork of bilateral and multilateral agreements, emanating from the Arctic Council and anchored firmly in UNCLOS. By reaching an enduring modus vivendi, the Arctic powers have set the stage for a long-lasting regional boom.

## A Region of Riches

Most cartographic depictions conceal the Arctic's physical vastness. Alaska, which US maps usually relegate to a box off the coast of California, is actually two and a half times as large as Texas and has more coastline than the lower 48 states combined. Greenland is larger than all of western Europe. The area inside the Arctic Circle contains eight percent of the earth's surface and 15 percent of its land.

It also includes massive oil and gas deposits—the main reason the region is so economically promising. Located primarily in western Siberia and Alaska's Prudhoe Bay, the Arctic's oil and gas fields account for 10.5 percent of global oil production and 25.5 percent of global gas production. And those numbers could soon jump. Initial estimates suggest that the Arctic may be home to an estimated 22 percent of the world's undiscovered conventional oil and gas deposits, according to the US Geological Survey. These riches have become newly accessible and attractive, thanks to retreating sea ice, a lengthening summer drilling season, and new exploration technologies.

Private companies are already moving in. Despite high extraction costs and regulatory hurdles, Shell has invested $5 billion to look for oil in Alaska's Chukchi Sea, and the Scottish company Cairn Energy has invested $1 billion to do the same off the coast of Greenland. Gazprom and Rosneft are planning to invest many billions of dollars more to develop the Russian Arctic, where the state-owned companies are partnering with ConocoPhillips, ExxonMobil, Eni, and Statoil to tap remote reserves in Siberia. The fracking boom may eventually exert downward pressure on oil prices, but it hasn't changed the fact that the Arctic contains tens of billions of barrels of conventional oil that will one day contribute to a greater global supply. Moreover, that boom has also reached the Arctic. Oil fracking exploration has already begun in northern Alaska, and this past spring, Shell and Gazprom signed a major deal to develop shale oil in the Russian Arctic.

Then there are the minerals. Now, longer summers are providing additional time to prospect mineral deposits, and retreating sea ice is opening deep-water ports for their export. The Arctic is already home to the world's most productive zinc mine, Red Dog, in northern Alaska, and its most productive nickel mine, in Norilsk, in northern Russia. Thanks mostly to Russia, the Arctic produces 40 percent of the world's palladium, 20 percent of its diamonds, 15 percent of its platinum, 11 percent of its cobalt, ten percent of its nickel, nine percent of its tungsten, and eight percent of its zinc. Alaska has more than 150 prospective deposits of rare-earth elements, and if the state were its own country, it would rank in the top ten in global reserves for many of these minerals. And all these assets are just the beginning. The Arctic has only begun to be surveyed. Once the digging starts, there is every reason to expect that, as often happens, even greater quantities of riches will be uncovered.

The coming Arctic boom will involve more than just mining and drilling. The region's Boreal forests of spruces, pines, and firs account for eight percent of the earth's total wood reserves, and its waters already produce ten percent of the world's total fishing catch. Converted tankers may someday ship clean water from Alaskan glaciers to southern Asia and Africa.

The Arctic's unique geography is an asset unto itself. Viewed from the top of the globe, the region sits at the crossroads of the world's most productive economies; Icelandair has started offering circum-polar service between Reykjavik, Anchorage, and St. Petersburg, and planned underwater telecommunications cables will link Northeast Asia, the northeastern United States, and Europe. The Arctic's high latitudes make the region a good place to expand existing ground stations for satellites in polar orbits. With some of the world's most powerful tides, the Arctic has spectacular hydropower potential, and its geology holds tremendous capacity for geothermal energy, as evidenced by Iceland's geothermal-powered aluminum smelting industry. Cool temperatures also make the Arctic an attractive place to construct data-storage centers, like the one Facebook is building in northern Sweden. A vault dug into the cool bedrock of the Svalbard Islands stores hundreds of thousands of plant seeds for preservation.

As the sea ice melts, once-fabled shipping shortcuts are becoming a reality. The Northwest Passage, which runs through the Canadian archipelago, remains choked with ice. But in 2010, for the first time in recorded history, commercial vessels— four of them—sailed from northwestern Europe to Northeast Asia via the Northern Sea Route, which passes through the Arctic Ocean above Eurasia. That number jumped to 34 in 2011 and to 46 during last year's Arctic summer. Although the Northern Sea Route has a long way to go before it siphons off a meaningful portion of traffic from the Suez and Panama Canals, it is no longer just a mariner's fantasy; it is an increasingly viable seaway for tankers looking to shave thousands of nautical miles off the traditional routes that go through the Strait of Malacca and

> *But a funny thing happened on the way to Arctic anarchy. Rather than harden positions, the possibility of increased tensions has spurred the countries concerned to work out their differences peacefully.*

the Strait of Gibraltar. It also provides a new export channel for warming farmlands and emerging mines along Russia's northern coast, where some of the country's largest rivers empty into the Arctic Ocean. Recognizing the route's promise, the Russian Ministry of Transport recently established a Northern Sea Route office in Moscow to handle shipping permits, monitor marine weather, and install new navigational aides along the passage. As the sea ice melts further, a route passing directly over the North Pole and avoiding the Russian coast altogether will also open.

## Fiscally Fit

Of course, natural resources and favorable geography alone aren't enough to make a region economically viable: just consider the Middle East. But the Arctic has much more going for it. For one thing, most of the countries with territory above the Arctic Circle are in relatively good fiscal shape. Denmark, Norway, Finland, and Sweden all have debt-to-GDP ratios under 54 percent, and Russia's is just 12 percent. Although the United States holds 75 percent of its GDP in debt, it has so far been shielded from high interest rates due to the dollar's role as a global currency reserve, and Alaska, for its part, runs budget surpluses and has earned an AAA credit rating from Standard & Poor's. At 84 percent, Canada's debt-to-GDP ratio is higher, but the country is exceedingly stable, with the World Economic Forum in 2012 ranking the Canadian banking system as the soundest in the world for the fifth consecutive year. Iceland is still grappling with the fallout from the 2008 collapse of its financial system, but it is recovering at record speed. In 2012, the country's GDP rose by 2.7 percent and unemployment declined to 5.6 percent. The general fiscal health of the Arctic countries means the region boasts an attractive environment for private capital, especially in comparison with other resource-rich frontiers.

Several Arctic countries possess sizable sovereign wealth funds, financed by royalties from their oil and gas production, resources that they can use to help finance infrastructure projects to stimulate development in the region. At more than $700 billion, Norway's sovereign wealth fund is the world's largest. Russia's National Welfare Fund stands at $175 billion. Alaska's Permanent Fund is worth $45 billion and allows the state to charge no income taxes; it even hands out an annual dividend to every resident. If governments think strategically, these reserves could fund the transportation and energy skeletons around which the Arctic's growing economy could mature.

With the glaring exception of Russia, the Arctic countries also boast predictable legal systems and clear regulations that are conducive to investment. The United States, Denmark, Norway, Iceland, Finland, Sweden, and Canada all rank in the top

20 on the World Bank's Ease of Doing Business Index. Thanks to the legal certainty provided by strong state institutions, these countries have little trouble attracting foreign capital; unlike with other frontier economies, investors can be pretty confident that the North American and Nordic governments will not nationalize private assets, demand kickbacks, or issue arbitrary court rulings.

## Bounty Hunting

No region so rich in resources, both real and man-made, can avoid attracting the attention of China for long. Indeed, right on cue, Beijing has begun a concerted effort to make inroads in the Arctic—especially in Iceland and its semiautonomous neighbor, Greenland—with far-reaching geopolitical implications. In May, the Arctic Council granted observer status to China, along with India, Italy, Japan, Singapore, and South Korea.

China sees Iceland as a strategic gateway to the region, which is why Premier Wen Jiabao made an official visit there last year (before heading to Copenhagen to discuss Greenland). China's state-owned shipping company is eyeing a long-term lease in Reykjavik, and the Chinese billionaire Huang Nubo has been trying for years to develop a 100-square-mile plot of land on the north of the island. In April, Iceland signed a free-trade deal with China, making it the first European country to do so. Whereas the United States closed its Cold War–era military base in Iceland in 2006, China is expanding its presence there, constructing the largest embassy by far in the country, sending in a constant stream of businesspeople, and dispatching its official icebreaker, the Xue Long, or "Snow Dragon," to dock in Reykjavik last August.

Greenland's main attraction, meanwhile, lies underground. In addition to iron ore and oil, the island sits on top of massive deposits of rare-earth elements, the global supply of which China dominates. Greenland may be home to fewer than 60,000 people, but the island has hosted a number of Asian delegations over the past few years. Last September, then South Korean President Lee Myung-bak came to attend the signing of an agreement between a South Korean state-owned mining company and a Greenlandic one. He was preceded by China's then minister of land and resources, Xu Shaoshi, who was also there to sign cooperation agreements. So far, these joint ventures have mostly been exploratory in nature, but they may soon yield megaprojects that feed resource-hungry markets in Asia.

Ever since Denmark granted it home rule in 1979, Greenland has been moving down the path to full autonomy, and in 2009, it gained control over its judicial system and natural resources. The local government has used that freedom to strike up commercial relationships with China, South Korea, and other countries. If foreign investment on the island continues at its current pace, local revenues could someday displace the $600 million annual subsidy that Greenland receives from Copenhagen, which could enable Greenland to demand political independence. Voters on the island endorsed this trajectory in March, when the pro-development Siumut Party won a plurality in Greenland's parliament. While equatorial microstates may soon disappear into the rising sea, Greenland might well become the first country born from climate change.

Meanwhile, the Arctic countries have been investing in their own icy backyards. Russia has led the way with enthusiastic presidential leadership and a number of state programs committed to spurring infrastructure investment along its northern coast. In Canada, the governments of the Yukon Territory, the Northwest Territories, Nunavut, and Quebec have established development offices to attract investment. In May, when Canada assumed the chairmanship of the Arctic Council, it appointed the head of its Northern Economic Development Agency as the country's senior Arctic official, instructing him to steer an Arctic Council policy of "development for the people of the north." For years, Norwegian companies have been launching joint ventures with their Russian counterparts to develop oil and gas projects around the Barents Sea. Alaska, meanwhile, has enacted pro-growth policies, such as lowering oil and gas taxes and selling more leases on state lands.

Yet Juneau has struggled in the face of obstructionism on the part of the federal government, which has kept federal lands closed and forces developers to navigate a burdensome permitting process and endure unending regulatory uncertainty. At this point, Alaska's leaders would prefer that the federal government just get out of the way. Washington's unhelpful attitude epitomizes its generally passive Arctic policy. While the rest of the world has already awoken to the region's growing importance, the United States still seems fast asleep, leaving the playing field open to more competitive rivals.

## Arctic Awakening

The good news is that it's not too late to play catch-up. The first and most obvious place for the United States to start is to finally join the 164 other countries that have acceded to UNCLOS. Ironically, Washington had a hand in drafting the original treaty, but Senate Republicans, making misguided arguments about the supposed threat the treaty poses to US sovereignty, have managed to block its ratification for decades. The result has been real harm to the national interest.

UNCLOS allows countries to claim exclusive jurisdiction over the portions of their continental shelves that extend beyond the 200-nautical-mile exclusive economic zones prescribed by the treaty. In the United States' case, this means that the country would gain special rights over an extra 350,000 square miles of ocean—an area roughly half the size of the entire Louisiana Purchase. Because the country is not a party to UNCLOS, however, its claims to the extended continental shelf in the Beaufort and Chukchi seas (and elsewhere) cannot be recognized by other states, and the lack of a clear legal title has discouraged private firms from exploring for oil and gas or mining the deep seabed. The failure to ratify UNCLOS has also relegated the United States to the back row when it comes to establishing new rules for the Arctic. Just as traffic through the Bering Strait is growing, Washington lacks the best tool to influence regulations governing sea-lanes and protecting fisheries and sensitive habitats. The treaty also enshrines the international legal principle of freedom of navigation, which the US Navy relies on to project power globally.

No wonder everyone from the head of the US Chamber of Commerce to the president of the Natural Resources Defense Council to the chairman of the Joint

Chiefs of Staff (along with every living secretary of state) has argued that the United States should ratify UNCLOS. It is far past time for the Senate to follow their advice. Skeptical Senate Republicans have stood in the way of ratification, arguing that the treaty would place limits on US sovereignty. But that argument is a red herring, since the United States already follows all of the treaty's guidelines anyway, and ratifying it would in fact give Washington new rights and greater influence. There are probably enough votes from moderate Republicans for the treaty to pass, if the president decided to make ratification a priority.

More broadly, Washington needs to continue developing a coherent approach to the Arctic, as other countries have already done. This May, the White House published the National Strategy for the Arctic Region. The document is a promising start, and it goes a long way toward updating the thin National Security Presidential Directive that the George W. Bush administration issued in 2009. In devising the strategy, the Obama administration deserves credit for reaching out to the Alaskan government, and especially to indigenous populations, whose voice and experience are critical. But the United States is late to the game, and there is still much work to be done in thinking through a national approach to the Arctic and developing the capabilities to project power there.

For starters, the United States needs to increase its presence in the Arctic. That, in turn, will require building icebreakers, since none of the US Navy's current surface ships are powerful enough to navigate in the Arctic. Here again, the United States lags behind its Arctic neighbors: Russia owns 30 icebreakers, some of them nuclear-powered, and Canada has 13. Even South Korea and China, which lack Arctic coastlines, own new icebreakers. The US Coast Guard has only three: one is inoperative, one was commissioned in 1976 and is on its last legs, and one is more of a floating research lab than a military tool.

Yet even if Congress appropriated the money to build icebreakers tomorrow, due to the Merchant Marine Act of 1920 (also known as the Jones Act), which requires ships traveling between US ports to be built in the United States, the Coast Guard estimates that it would take a decade for the United States' moribund shipyards to construct a single new vessel—by which time the Arctic summer sea ice would likely have already disappeared. Congress should relax this protectionist law to allow the Coast Guard and the navy to procure foreign-built ships or lease privately built American ones, at a fraction of the cost.

The United States also has no Arctic deep-water port, no military aviation facility in the region, and no comprehensive network for monitoring Arctic shipping, which would prove especially useful in the Bering Strait, the 55-mile-wide chokepoint between the Pacific and Arctic oceans. The federal government should build on the real progress that Alaska has made in these areas on its own over the last few years. Washington need not spend as much as it did building the canals, bridges, dams, and roads that opened up the American West, but some minimum investments would help the United States compete in the region.

Finally, the United States needs to reinvigorate its Arctic diplomacy. Following in the footsteps of other countries in the region (as well as Japan and Singapore), it should appoint a high-level diplomat as an Arctic ambassador to represent US interests in such forums as the Arctic Council. By dispatching junior diplomats to Arctic meetings where other countries are represented by their foreign ministers, as Washington sometimes does, the United States sends a clear message: that the region doesn't matter. In May, Secretary of State John Kerry attended an Arctic Council meeting, just as Secretary of State Hillary Clinton had done before him, and the practice should continue. To remind Americans that they live in an Arctic nation, President Barack Obama should highlight the Arctic in an address to Congress, the way Canadian Prime Minister Stephen Harper and Russian President Vladimir Putin have done before their legislatures.

More engagement could even improve US-Russian relations. According to the 1867 treaty by which the Russian empire sold Alaska to the United States, the two countries were "desirous of strengthening, if possible, the good understanding which exists between them," and then US Secretary of State William Seward hoped the purchase would do just that. Good relations have eluded the United States and Russia for many of the decades that followed, but today, the Arctic could become the source of the cooperation that Seward foresaw. In the Bering Sea, Russia and the United States possess common objectives, and there is ample room for cooperation on policing foreign fishing fleets, responding to oil spills, and aiding navigation.

## A New Kind of Development

Climate change is transforming the Arctic from a geopolitical afterthought into an epic bounty ripe for this century's entrepreneurs. Countries should continue their commitment to the peaceful course they have charted there so far. But policymakers need to get serious about establishing a shared vision of how to harness the Arctic's resources. Economic development need not mean environmental disaster. Indeed, the opening up of the Arctic offers a once-in-a-lifetime opportunity to develop a frontier economy sustainably.

For such an approach to catch on, countries will have to strike the right balance between environmentalism and exploitation. One way to blend capitalism with conservationism is to value nature as a form of capital and price the environment into development decisions, as programs that manage fisheries by allocating catch shares have done and as programs that protect forests by creating tradable securities have done, too. For this tactic to work in the Arctic, there needs to be a full accounting of the available resources, which is why it is so important for governments, nongovernmental organizations, and others to conduct a comprehensive census of the region's natural resources and biological diversity. As better scientific baselines are established, governments can make informed decisions about development, balancing the risks to this sensitive environment with their other economic and national security priorities. The goal should be to find a middle ground between the environmental activists who want to immediately turn the Arctic into a nature preserve and the "Drill, baby, drill!" crowd, which prizes resource exploitation above all else.

In Alaska, this means allowing oil and gas projects to proceed on a case-by-case basis but using some of the profits to create a more diversified economy. Otherwise, the state risks becoming just another petrocolony laid low by the resource curse. Alaska should invest its considerable wealth in its underdeveloped university system, finance ambitious infrastructure projects, and create policies that attract talented immigrants and encourage them to start new businesses, such as renewable energy ventures. The model to follow is Norway, which took advantage of an oil windfall to fund a progressive state and kick-start its renewable energy sector. Such an approach would be deeply Alaskan, too, consistent with the state constitution's order that Alaska "encourage the settlement of its land and the development of its resources by making them available for maximum use consistent with the public interest."

The Arctic presents an extraordinary opportunity to rewrite the rules of the game for developing a frontier economy. But the time to start doing so is now, before a Deepwater Horizon–like oil spill stains the Arctic and its appeal. With the Arctic heating up faster than many predicted, it is a matter of not if but when the summer sea ice will be gone and the region will open up to widespread development. If managed correctly, the Arctic could be both a carefully protected environment and a major driver of economic growth—with enormous benefits for both outsiders and the inhabitants of this prime real estate.

# 6

# Health and Human Factors

US Coast Guard/EPA/Landov

A fire aboard the mobile offshore drilling rig Deepwater Horizon in the Gulf of Mexico, April 22, 2010, in what is the worst oil spill in the history of the United States.

# What Happens to Us?

Increasing global temperatures occurring as a result of global warming are expected to have significant consequences for humankind. Many of the effects of warmer temperatures on the planet's landscape are predictable, though not with any precision. A rising average global temperature, for example, will certainly bring about an increase in ocean water temperature accompanied by thermal expansion and an increase in the rate of melt of long-standing ice cover. The level of the water in the oceans will inevitably rise. A rise in ocean water temperature will add more energy to storms, making them more violent and leading them to carrying more water inland in the form of rain or snowfall. Changes to local temperatures occurring as a result of increased global temperatures will cause weather patterns to change. The precise extent to which any of these events will occur is only marginally predictable. Regardless of the actual extent of increasing temperatures, what is a certainty is that the human population will be forced to contend with the consequences of a warmer world.

## Threats to Humanity Due to Climate Change

Most scientists believe that climate change is the direct result of excess carbon being emitted into the atmosphere from human industrial activity—specifically, the burning of fossil fuels such as oil, gas, and coal. The debate over the severity and causes of global warming is exacerbated by subjective interpretation of complex data. Nevertheless, that the world has experienced a widespread increase in average temperatures is undisputed.

In a 2009 report, the Global Humanitarian Forum referred to climate change as a "silent crisis." According to the report, millions of people worldwide are already suffering from the adverse effects of increasing global temperatures. These effects include increased threats to food supplies, health, freshwater sources, and security.

The issue of global warming's impact on daily life worldwide has two principal components. One is the actual physical effects of climate change, including disasters related to the weather, the process of desertification, and rising sea levels. These conditions are already negatively affecting communities all over the globe. The other component is the issue of social justice. It is widely agreed that the peoples and nations who are the least responsible for emitting the carbon dioxide associated with climate change are also those who are least able to defend against its effects. Research shows increasing temperatures have affected the monsoon season in Sudan, causing widespread drought in the country's Darfur region and putting millions of lives at risk. Residents of the island nation of Kiribati have been warned of a one- to three-foot rise in ocean levels, which would force the evacuation of an estimated 100,000 residents. Other nations that face forced evacuations as water levels increase include the Netherlands and Bangladesh.

The physical changes that are certain to take place through global warming will undoubtedly aggravate existing development challenges such as hunger, disease, and poverty. Global warming threatens economic growth and social stability throughout the world. Available scientific data does not allow for precise estimates of the impact of global warming on humanity. Moreover, there is as yet no clear consensus on the causal relationship between weather-related disasters and climate change; the multitude of other possible factors involved—natural variability of local climates, population growth, land use, and governance and social accommodation—make it difficult to isolate global climate change as the cause. For instance, events such as unnaturally powerful Atlantic hurricanes and Pacific cyclones can be accounted for as short-term localized phenomena. Nonetheless, climate change may in fact be a contributing factor in driving such storms.

Global warming is expected to have a cumulative effect on human society and is likely to aggravate existing global problems related to health and development.

## Specific Climate Change Effects and Human Problems

As the amount of carbon dioxide in the atmosphere continues to build, scientists expect the earth's average surface temperature and ocean water temperatures to increase slowly over time. The progression is essentially imperceptible. The average ocean temperature is believed to have increased by just 1 degree Fahrenheit in the past century. While this amount seems miniscule, this increase means that a huge amount of additional thermal energy has been absorbed by the earth's oceans. In the Arctic and Antarctic regions, seawater is in contact with ice and the water temperature remains fairly constant as long as the ice is present. In tropical and subtropical regions, seawater temperature can increase by several degrees above its historical norm. Circulation patterns also play a large role in locally changing water temperatures.

Rising seawater temperatures have a significant effect on the environment. Though the ice-seawater temperature remains almost unchanged, additional energy in the water is absorbed by the ice, causing it to melt at increased rates. Meltwater adds to the water volume of the oceans, raising the sea level at a rate that is imperceptible to populations living along the seaboard. The sea level has increased by twenty centimeters (eight inches) over the past century, or about two millimeters per year. Indications are that the rate of sea level rise is increasing. Warmer water, being less dense than water of a lower temperature, also occupies more volume, which enhances the rate of sea level rise. It is estimated that by the end of the twenty-first century, ocean water levels will have risen by one meter. For coastal populations, such a rise in water level will mean that huge areas of land that are densely populated will be rendered uninhabitable by salt-laden floodwaters. Examples of this effect have already been witnessed in the present day in New Orleans, Louisiana, in the Ganges Delta of India, and in Venice, Italy. Farther inland, adjacent agricultural areas will become subject to flooding by storm surges and be rendered agriculturally useless due to the deposition of salt from floodwaters. This has also already been observed in the Ganges Delta region.

Rising surface temperatures will exacerbate or intensify naturally occurring meteorological phenomenon in many regions. The combination of the earth's rotation and seasonal variation generates strong, moist updrafts near the equator. These winds rise and cool, producing rain and other precipitation in the subtropical regions of the Northern and Southern Hemispheres. The cold, dry air then descends, absorbing moisture and heat from the regions that have become characterized by this action as arid lands and deserts. The airflow then returns to the equatorial regions and rises again in a continuous cycle. A similar meteorological system drops rain and other precipitation in the northern and southern temperate zones of the planet before descending again in the polar desert regions. Increasing temperatures in the equatorial region exerts a number of effects. These include the expansion of tropical zones, low rainfall, and drought conditions. Warmer, moist air is driven farther afield, extending the area of desertification and pushing warmer climates farther north and south. Higher moisture in the temperate regions results in changes to local precipitation and run-off patterns. Unusually large snowfalls have occurred recently in North America and Europe, and unusually large amounts of rain have produced severe floods in normally semiarid regions such as Canada's southern Prairie Provinces and central Europe.

Expanding climatic zones have resulted in the expansion of ranges that are hospitable to specific animal species, introducing them into areas where they are not historically known. In addition, changing climate patterns have increased species extinction. Animal species adapt to specific environments in localized terrains. When environmental conditions change sufficiently, species that do not adapt are at a higher risk of extinction. Moving with the expanding climate zone also puts species at risk as they are forced to encroach on the territories of native species and compete for the resources in that area. As plant and animal species decline through this natural attrition, and as human activity encroaches on and destroys habitats, species diversity declines, further weakening ecosystems.

Species migration with expanded climatic zones also means certain bacteria and viruses that are inimical to humans can move into regions that had previously been free of their threat. Parasite-borne diseases that have until recently been indigenous to the tropics and subtropics are appearing in new regions. The Chikungunya virus, West Nile virus, and dengue fevers, for example, have all been reported in recent years as occurring in the Mediterranean area of Europe and in North America. Infectious diseases now found in areas outside their historical range may have been transferred through the effects of climate change. The frequency and relative ease of travel in the globalized economy also increases the transfer of communicable diseases from place to place. This spread of disease will likely place additional burden on medical infrastructure and governing bodies.

Worldwide, the effects of global warming are becoming increasingly evident. Its effects on human health are likely to become more acute in the future.

# Climate Change and Health

## World Health Organization, October 2012

## Key Facts

- Climate change affects the social and environmental determinants of health—clean air, safe drinking water, sufficient food and secure shelter.

- Global warming that has occurred since the 1970s caused over 140,000 excess deaths annually by the year 2004.

- The direct damage costs to health (i.e., excluding costs in health-determining sectors such as agriculture and water and sanitation), is estimated to be between US $2–4 billion/year by 2030.

- Many of the major killers such as diarrhoeal diseases, malnutrition, malaria and dengue are highly climate-sensitive and are expected to worsen as the climate changes.

- Areas with weak health infrastructure—mostly in developing countries—will be the least able to cope without assistance to prepare and respond.

- Reducing emissions of greenhouse gases through better transport, food and energy-use choices can result in improved health.

## Climate Change

Over the last 50 years, human activities—particularly the burning of fossil fuels—have released sufficient quantities of carbon dioxide and other greenhouse gases to trap additional heat in the lower atmosphere and affect the global climate.

In the last 100 years, the world has warmed by approximately 0.75°C. Over the last 25 years, the rate of global warming has accelerated, at over 0.18°C per decade.[1]

Sea levels are rising, glaciers are melting and precipitation patterns are changing. Extreme weather events are becoming more intense and frequent.

## What Is the Impact of Climate Change on Health?

Although global warming may bring some localized benefits, such as fewer winter deaths in temperate climates and increased food production in certain areas, the overall health effects of a changing climate are likely to be overwhelmingly negative. Climate change affects social determinants of health—clean air, safe drinking water, sufficient food and secure shelter.

## Extreme Heat

Extremely high air temperatures contribute directly to deaths from cardiovascular and respiratory disease, particularly among elderly people. In the heat wave of summer 2003 in Europe, for example, more than 70,000 excess deaths were recorded.[2] High temperatures also raise the levels of ozone and other pollutants in the air that exacerbate cardiovascular and respiratory disease. Urban air pollution causes about 1.2 million deaths every year.

Pollen and other aeroallergen levels are also higher in extreme heat. These can trigger asthma, which affects around 300 million people. Ongoing temperature increases are expected to increase this burden.

## Natural Disasters and Variable Rainfall Patterns

Globally, the number of reported weather-related natural disasters has more than tripled since the 1960s. Every year, these disasters result in over 60,000 deaths, mainly in developing countries.

Rising sea levels and increasingly extreme weather events will destroy homes, medical facilities and other essential services. More than half of the world's population lives within 60 km of the sea. People may be forced to move, which in turn heightens the risk of a range of health effects, from mental disorders to communicable diseases.

Increasingly variable rainfall patterns are likely to affect the supply of fresh water. A lack of safe water can compromise hygiene and increase the risk of diarrhoeal disease, which kills 2.2 million people every year. In extreme cases, water scarcity leads to drought and famine. By the 2090s, climate change is likely to widen the area affected by drought, double the frequency of extreme droughts and increase their average duration six-fold.[3]

Floods are also increasing in frequency and intensity. Floods contaminate freshwater supplies, heighten the risk of water-borne diseases, and create breeding grounds for disease-carrying insects such as mosquitoes. They also cause drownings and physical injuries, damage homes and disrupt the supply of medical and health services.

Rising temperatures and variable precipitation are likely to decrease the production of staple foods in many of the poorest regions—by up to 50 percent by 2020 in some African countries.[4] This will increase the prevalence of malnutrition and undernutrition, which currently cause 3.5 million deaths every year.

## Patterns of Infection

Climatic conditions strongly affect water-borne diseases and diseases transmitted through insects, snails or other cold blooded animals.

Changes in climate are likely to lengthen the transmission seasons of important vector-borne diseases and to alter their geographic range. For example, climate change is projected to widen significantly the area of China where the snail-borne disease schistosomiasis occurs.[5]

Malaria is strongly influenced by climate. Transmitted by Anopheles mosquitoes, malaria kills almost 1 million people every year—mainly African children under five years old. The Aedes mosquito vector of dengue is also highly sensitive to climate conditions. Studies suggest that climate change could expose an additional 2 billion people to dengue transmission by the 2080s.[6]

## Measuring the Health Effects

Measuring the health effects from climate change can only be very approximate. Nevertheless, a WHO assessment, taking into account only a subset of the possible health impacts, concluded that the modest warming that has occurred since the 1970s was already causing over 140,000 excess deaths annually by the year 2004.[7]

## Who Is at Risk?

All populations will be affected by climate change, but some are more vulnerable than others. People living in small island developing states and other coastal regions, megacities, and mountainous and polar regions are particularly vulnerable.

> *Globally, the number of reported weather-related natural disasters has more than tripled since the 1960s. Every year, these disasters result in over 60,000 deaths, mainly in developing countries.*

Children—in particular, children living in poor countries—are among the most vulnerable to the resulting health risks and will be exposed longer to the health consequences. The health effects are also expected to be more severe for elderly people and people with infirmities or pre-existing medical conditions.

Areas with weak health infrastructure—mostly in developing countries—will be the least able to cope without assistance to prepare and respond.

## WHO Response

Many policies and individual choices have the potential to reduce greenhouse gas emissions and produce major health co-benefits. For example, promoting the safe use of public transportation and active movement—such as cycling or walking as alternatives to using private vehicles—could reduce carbon dioxide emissions and improve health.

In 2009, the World Health Assembly endorsed a new WHO workplan on climate change and health. This includes:

- **Advocacy:** to raise awareness that climate change is a fundamental threat to human health.

- **Partnerships:** to coordinate with partner agencies within the UN system, and ensure that health is properly represented in the climate change agenda.

- **Science and evidence:** to coordinate reviews of the scientific evidence on the links between climate change and health, and develop a global research agenda.

- **Health system strengthening:** to assist countries to assess their health vulnerabilities and build capacity to reduce health vulnerability to climate change.

## References

1.  Based on data from the United Kingdom Government Met Office. *HadCRUT3 annual time series,* Hadley Research Centre, 2008.
2.  Robine JM et al. Death toll exceeded 70,000 in Europe during the summer of 2003. *Les Comptes Rendus/Série Biologies,* 2008, 331:171–78.
3.  Arnell NW. Climate change and global water resources: SRES emissions and socio-economic scenarios. *Global Environmental Change—Human and Policy Dimensions,* 2004, 14:31–52.
4.  *Climate change 2007. Impacts, adaptation and vulnerability.* Geneva, Intergovernmental Panel on Climate Change, 2007 (Contribution of Working Group II to the Fourth Assessment Report of the Intergovernmental Panel on Climate Change).
5.  Zhou XN et al. Potential impact of climate change on schistosomiasis transmission in China. *American Journal of Tropical Medicine and Hygiene,* 2008, 78:188–194.
6.  Hales S et al. Potential effect of population and climate changes on global distribution of dengue fever: an empirical model. *The Lancet,* 2002, 360:830–834.
7.  *Global health risks: mortality and burden of disease attributable to selected major risks.* World Health Organization, Geneva, 2009.

# How We Know Global Warming Is Real and Human Caused

By Tapio Schneider
*Skeptic,* 2012

## The Science behind Human-Induced Climate Change

Atmospheric carbon dioxide concentrations are higher today than at any time in at least the past 650,000 years. They are about 35 percent higher than before the industrial revolution, and this increase is caused by human activities, primarily the burning of fossil fuels. Carbon dioxide is a greenhouse gas, as are methane, nitrous oxide, water vapor, and a host of other trace gases. They occur naturally in the atmosphere. Greenhouse gases act like a blanket for infrared radiation, retaining radiative energy near the surface that would otherwise escape directly to space. An increase in atmospheric concentrations of carbon dioxide and of other greenhouse gases augments the natural greenhouse effect; it increases the radiative energy available to Earth's surface and to the lower atmosphere. Unless compensated for by other processes, the increase in radiative energy available to the surface and the lower atmosphere leads to warming. This we know. How do we know it?

## How Do We Know Carbon Dioxide Concentrations Have Increased?

The concentrations of carbon dioxide and other greenhouse gases in atmospheric samples have been measured continuously since the late 1950s. Since then, carbon dioxide concentrations have increased steadily from about 315 parts per million (ppm, or molecules of carbon dioxide per million molecules of dry air) in the late 1950s to about 385 ppm now, with small spatial variations away from major sources of emissions. For the more distant past, we can measure atmospheric concentrations of greenhouse gases in bubbles of ancient air preserved in ice (e.g., in Greenland and Antarctica). Ice core records currently go back 650,000 years; over this period we know that carbon dioxide concentrations have never been higher than they are now. Before the industrial revolution, they were about 280 ppm, and they have varied naturally between about 180 ppm during ice ages and 300 ppm during warm periods. Concentrations of methane and nitrous oxide have likewise increased since the industrial revolution and, for methane, are higher now than they have been in the 650,000 years before the industrial revolution.

---

## How Do We Know the Increase in Carbon Dioxide Concentrations Is Caused by Human Activities?

There are several lines of evidence. We know approximately how much carbon dioxide is emitted as a result of human activities. Adding up the human sources of carbon dioxide—primarily from fossil fuel burning, cement production, and land use changes (e.g., deforestation)—one finds that only about half the carbon dioxide emitted as a result of human activities has led to an increase in atmospheric concentrations. The other half of the emitted carbon dioxide has been taken up by oceans and the biosphere—where and how exactly is not completely understood: there is a "missing carbon sink."

Human activities thus can account for the increase in carbon dioxide concentrations. Changes in the isotopic composition of carbon dioxide show that the carbon in the added carbon dioxide derives largely from plant materials, that is, from processes such as burning of biomass or fossil fuels, which are derived from fossil plant materials. Minute changes in the atmospheric concentration of oxygen show that the added carbon dioxide derives from burning of the plant materials. And concentrations of carbon dioxide in the ocean have increased along with the atmospheric concentrations, showing that the increase in atmospheric carbon dioxide concentrations cannot be a result of release from the oceans. All lines of evidence taken together make it unambiguous that the increase in atmospheric carbon dioxide concentrations is human induced and is primarily a result of fossil fuel burning. (Similar reasoning can be evoked for other greenhouse gases, but for some of those, such as methane and nitrous oxide, their sources are not as clear as those of carbon dioxide.)

## How Can Such a Minute Amount of Carbon Dioxide Affect Earth's Radiative Energy Balance?

Concentrations of carbon dioxide are measured in parts per million, those of methane and nitrous oxide in parts per billion. These are trace constituents of the atmosphere. Together with water vapor, they account for less than 1 percent of the volume of the atmosphere. And yet they are crucially important for Earth's climate.

Earth's surface is heated by absorption of solar (shortwave) radiation; it emits infrared (longwave) radiation, which would escape almost directly to space if it were not for water vapor and the other greenhouse gases. Nitrogen and oxygen, which account for about 99 percent of the volume of the atmosphere, are essentially transparent to infrared radiation. But greenhouse gases absorb infrared radiation and re-emit it in all directions. Some of the infrared radiation that would otherwise directly escape to space is emitted back toward the surface. Without this natural greenhouse effect, primarily owing to water vapor and carbon dioxide, Earth's mean surface temperature would be a freezing -1°F, instead of the habitable 59°F we currently enjoy. Despite their small amounts, then, the greenhouse gases strongly affect Earth's temperature. Increasing their concentration augments the natural greenhouse effect.

## How Do Increases in Greenhouse Gas Concentrations Lead to Surface Temperature Increases?

Increasing the concentration of greenhouse gases increases the atmosphere's "optical thickness" for infrared radiation, which means that more of the radiation that eventually does escape to space comes from higher levels in the atmosphere. The mean temperature at the level from which the infrared radiation effectively escapes to space (the emission level) is determined by the total amount of solar radiation absorbed by Earth. The same amount of energy Earth receives as solar radiation, in a steady state, must be returned as infrared radiation; the energy of radiation depends on the temperature at which it is emitted and thus determines the mean temperature at the emission level. For Earth, this temperature is -1°F—the mean temperature of the surface if the atmosphere would not absorb infrared radiation. Now, increasing greenhouse gas concentrations implies raising the emission level at which, in the mean, this temperature is attained. If the temperature decreases between the surface and this level and its rate of decrease with height does not change substantially, then the surface temperature must increase as the emission level is raised. This is the greenhouse effect. It is also the reason that clear summer nights in deserts, under a dry atmosphere, are colder than cloudy summer nights on the US East Coast, under a relatively moist atmosphere.

In fact, Earth's surface temperatures have increased by about 1.3°F over the past century. The temperature increase has been particularly pronounced in the past 20 years. The scientific consensus about the cause of the recent warming was summarized by the Intergovernmental Panel on Climate Change (IPCC) in 2007: "Most of the observed increase in global average temperatures since the mid-20th century is very likely due to the observed increase in anthropogenic greenhouse gas concentrations. . . . The observed widespread warming of the atmosphere and ocean, together with ice mass loss, support the conclusion that it is extremely unlikely that global climate change of the past 50 years can be explained without external forcing, and very likely that it is not due to known natural causes alone."

The IPCC conclusions rely on climate simulations with computer models. Based on spectroscopic measurements of the optical properties of greenhouse gases, we can calculate relatively accurately the impact increasing concentrations of greenhouse gases have on Earth's radiative energy balance. For example, the radiative forcing owing to increases in the concentrations of carbon dioxide, methane, and nitrous oxide in the industrial era is about 2.3 Watts per square meter. (This is the change in radiative energy fluxes in the lower troposphere before temperatures have adjusted.) We need computer models to translate changes in the radiative energy balance into changes in temperature and other climate variables because feedbacks in the climate system render the climate response to changes in the atmospheric composition complex, and because other human emissions (smog) also affect climate in complex ways. For example, as the surface and lower atmosphere warm in response to increases in carbon dioxide concentrations, the atmospheric concentration of water vapor near the surface increases as well. That this has to happen is well established on the basis of the energy balance of the surface and relations between evaporation rates and the relative humidity of the atmosphere (it is not directly, as is sometimes stated, a consequence of higher evaporation rates).

Water vapor, however, is a greenhouse gas in itself, and so it amplifies the temperature response to increases in carbon dioxide concentrations and leads to greater surface warming than would occur in the absence of water vapor feedback. Other feedbacks that must be taken into account in simulating the climate response to changes in atmospheric composition involve, for example, changes in cloud cover, dynamical changes that affect the rate at which temperature decreases with height and hence affect the strength of the greenhouse effect, and surface changes (e.g., loss of sea ice). Current climate models, with Newton's laws of motion and the laws of thermodynamics and radiative transfer at their core, take such processes into account. They are able to reproduce, for example, Earth's seasonal cycle if all such processes are taken into account but not, for example, if water vapor feedback is neglected. The IPCC's conclusion is based on the fact that these models can only match the observed climate record of the past 50 years if they take human-induced changes in atmospheric composition into account. They fail to match the observed record if they only model natural variability, which may include, for example, climate responses to fluctuations in solar radiation.

Climate feedbacks are the central source of scientific (as opposed to socioeconomic) uncertainty in climate projections. The dominant source of uncertainty is cloud feedbacks, which are incompletely understood. The area covered by low stratus clouds may increase or decrease as the climate warms. Because stratus clouds are low, they do not have a strong greenhouse effect (the strength of the greenhouse effect depends on the temperature difference between the surface and the level from which infrared radiation is emitted, and this is small for low clouds); however, they reflect sunlight, and so exert a cooling effect on the surface, as anyone knows who has been near southern California's coast on an overcast spring morning. If their area coverage increases as greenhouse gas concentrations increase, the surface temperature response will be muted; if their area coverage decreases, the surface temperature response will be amplified. It is currently unclear how these clouds respond to climate change, and climate models simulate widely varying responses. Other major uncertainties include the effects of aerosols (smog) on clouds and the radiative balance and, on timescales longer than a few decades, the response of ice sheets to changes in temperature.

Uncertainties notwithstanding, it is clear that increases in greenhouse gas concentrations, in the global mean, will lead to warming. Although climate models differ in the amount of warming they project, in its spatial distribution, and in other more detailed aspects of the climate response, all climate models that can reproduce observed characteristics such as the seasonal cycle project warming in response to the increases in greenhouse gas concentrations that are expected in the coming decades as a result of continued burning of fossil fuels and other human activities such as tropical deforestation. The projected consequences of the increased concentrations of greenhouse gases have been widely publicized. Global-mean surface temperatures are likely to increase by 2.0 to 11.5°F by the

> *Increasing the concentration of greenhouse gases increases the atmosphere's "optical thickness" for infrared radiation, which means that more of the radiation that eventually does escape to space comes from higher levels in the atmosphere.*

year 2100, with the uncertainty range reflecting scientific uncertainties (primarily about clouds) as well as socio-economic uncertainties (primarily about the rate of emission of greenhouse gases over the 21st century). Land areas are projected to warm faster than ocean areas. The risk of summer droughts in mid-continental regions is likely to increase. Sea level is projected to rise, both by thermal expansion of the warming oceans and by melting of land ice.

Less widely publicized but important for policy considerations are projected very long-term climate changes, of which some already now are unavoidable. Even if we were able to keep the atmospheric greenhouse gas concentration fixed at its present level—this would require an immediate and unrealistically drastic reduction in emissions—the Earth's surface would likely warm by another 0.9–2.5°F over the next centuries. The oceans with their large thermal and dynamic inertia provide a buffer that delays the response of the surface climate to changes in greenhouse gas concentrations. The oceans will continue to warm over about 500 years. Their waters will expand as they warm, causing sea level rise. Ice sheets are thought to respond over timescales of centuries, though this is challenged by recent data from Greenland and Antarctica, which show evidence of a more rapid, though possibly transient, response. Their full contribution to sea level rise will take centuries to manifest. Studies of climate change abatement policies typically end in the year 2100 and thus do not take into account that most of the sea level rise due to the emission of greenhouse gases in the next 100 years will occur decades and centuries later. Sea level is projected to rise 0.2–0.6 meters by the year 2100, primarily as a result of thermal expansion of the oceans; however, it may eventually reach values up to several meters higher than today when the disintegration of glaciers and ice sheets contributes more strongly to sea level rise. (A sea level rise of 4 meters would submerge much of southern Florida.)

## Certainties and Uncertainties

While there are uncertainties in climate projections, it is important to realize that the climate projections are based on sound scientific principles, such as the laws of thermodynamics and radiative transfer, with measurements of optical properties of gases. The record of past climate changes that can be inferred, for example, with geochemical methods from ice cores and ocean sediment cores, provides tantalizing hints of large climate changes that occurred over Earth's history, and it poses challenges to our understanding of climate (for example, there is no complete and commonly accepted explanation for the cycle of ice ages and warm periods). However, climate models are not empirical, based on correlations in such records, but incorporate our best understanding of the physical, chemical, and biological processes being modeled. Hence, evidence that temperature changes precede changes in carbon dioxide concentrations in some climate changes on the timescales of ice ages, for example, only shows that temperature changes can affect the atmospheric carbon dioxide concentrations, which in turn feedback on temperature changes. Such evidence does not invalidate the laws of thermodynamics and radiative transfer, or the conclusion that the increase in greenhouse gas concentrations in the past decades is human induced.

# Soot Is No. 2 Global-Warming Culprit, Study Finds

By Pete Spotts
*Christian Science Monitor,* January 15, 2013

*From diesel engines to cow-dung cook fires, soot from inefficiently burned fuel has sup-planted methane as the second most significant global-warming agent that humans are pumping into the air, according to an exhaustive review of more than a decade's worth of research on black-carbon soot emissions.*

Carbon dioxide from burning fossil fuel and from land-use changes remains in the No. 1 spot. But the direct effect of soot on air temperatures, as well as its indirect effect on ice and snow melt and on cloud formation and persistence, are knocking at the door.

Given the uncertainties in the estimates, black-carbon soot may even outpace $CO_2$'s warming effect, according to the 232-page study published today in the *Journal of Geophysical Research–Atmospheres.*

Soot remains in the atmosphere for around seven days—a far shorter time than $CO_2$, which remains in the atmosphere for centuries. This means efforts to reduce soot may apply an important brake to warming in the short term with quick results, the researchers suggest.

Over the long term, however, countries still will have to solve the vexing political and economic challenges of tackling $CO_2$ emissions.

"There's a lot of promise in reducing black carbon" and other relatively short-lived warming agents, such as methane, says Tami Bond, an associate professor of civil and environmental engineering at the University of Illinois at Urbana-Champaign and one of the study's three lead authors. "But there's also a lot of caution."

Properly done, moving to reduce black-carbon soot has immediate climate and public-health benefits, she says. But the uncertainties surrounding some of its climate effects remain large.

Rather than serving as an excuse for inaction, however, the uncertainties should to serve as a guide for research, she adds.

For example, the processes that produce soot also produce not only $CO_2$ emissions but also other particles that can cool the atmosphere, she notes. In other words, emissions from any one source may contain competing influences on global warming.

The fingerprints "of human actions on climate are more complex than just the $CO_2$ story," she says.

Concern over the climate effects of black-carbon soot date back to at least 1971, when interest began to grow in the role small particles, or aerosols, could play in Earth's climate system.

During the past decade or so, however, field studies of soot's effect as a climate warmer typically yielded estimates two to three times higher than the effect seen in climate models, notes Veerabhadran Ramanathan, a professor of atmospheric and climate science at the Scripps Institution of Oceanography in La Jolla, Calif.

With this new study, "we're coming closer to what we think black carbon is doing to the planet's climate," says Dr. Ramanathan, who was not a member of the team that pulled together the new analysis but has been studying the impact of soot on the climate for much of his career.

The soot comes from a mix of sources that varies by region. The study notes that roughly 90 percent of global soot emissions fall into several broad categories: diesel-fueled vehicles; use of coal to heat or cook in homes; small kilns and industrial boilers; burning wood or other biomass for cooking; and open burning of biomass, such as using fires to clear forests for farming.

A study published last year suggested that the use of kerosene lamps was also a significant source. Soot from the lamps contributes about 270 billion tons of soot a year to the atmosphere, representing about 7 percent of warming impact from all energy-related soot, according to the study published last November in the journal Environmental Science and Technology.

The new study estimates that in 2000, humans injected soot into the air at a pace of about 7.5 trillion tons a year globally.

Researchers describe the warming effect in terms of the amount of energy deposited on a patch of Earth one meter square. As of 2005, the study notes, the accumulated direct and indirect effects of carbon-dioxide emissions since the dawn of the Industrial Revolution amounted to about 1.56 watts per square meter. Methane was No. 2 at 0.86 watts per square meter. The latest estimate for black-carbon soot now puts it's contribution to the energy warming the planet at 1.1 watts per square meter and perhaps as much as 2.1 watts per square meter.

The effects vary by region as well. Eastern and southern Asia, major sources for the soot, can experience a warming effect from soot 10 times higher than the global average, the researchers estimate. Soot falling on ice and snow in the Arctic has accelerated warming there by speeding the pace at which snow and ice melt during the long hours of summer sun.

Indeed, the mid to high latitudes of the northern hemisphere have seen some of the most pronounced effects of warming from soot, the study notes, because that's where most of the world's population lives.

Scientists also have linked high soot levels to shifts in the regional distribution and intensity of rainfall during Asian monsoons.

Initially, calls from some climate scientists during the past decade to focus first on reducing emissions of shorter-lived but powerful warming agents, such as

> *Soot from the lamps contributes about 270 billion tons of soot a year to the atmosphere, representing about 7 percent of warming impact from all energy-related soot. . . .*

methane or soot, have been met with polite nods and a resumption of heated debates over $CO_2$ emissions. But the increasing recognition of the adverse health effects of soot, as well as the experience from efforts to control soot, are changing that, some researchers say.

Indeed, "there are clear options" to cut soot emissions, Ramanathan says.

In California, for instance, black-carbon soot emissions fell by half between 1990 to 2008 in response to tighter air-quality regulations affecting diesel emissions, according to a study Ramanathan and colleagues from Scripps and Argonne National Laboratory published in early 2011.

The decline occurred even as "diesel consumption has increased significantly," he adds. The soot pollution "has come down to almost nothing" statewide.

The study on global black-carbon soot released [January 15, 2013] notes that focusing initially on diesel sources "appears to offer the most confidence in reducing near-term" warming.

Another opportunity lies in supplying cook stoves that burn biofuels like wood or dung more efficiently, Ramanathan adds.

In the end, putting an immediate focus on short-lived warming agents such as black carbon also cuts into $CO_2$ emissions, notes Brenda Ekwurzel, a climate scientist with the Union of Concerned Scientists in Washington.

"Emitting black carbon also emits carbon dioxide, which has longer-term consequences," she says. "So going after black carbon also helps with the carbon-dioxide emissions associated with incomplete combustion of an ancient tree or critter."

# Smoke Jumpers

By Elizabeth Grossman
*Earth Island Journal,* Summer 2012

*With global action on reducing $CO_2$ emissions all but stalled, governments focus their energy on another global warming pollutant: black carbon.*

International negotiations to reduce carbon dioxide emissions and slow global warming are stuck in a stalemate. Many people in the United States—the world's biggest economy and one of the planet's top per capita greenhouse gas emitters—continue to doubt the reality of manmade climate change. After a brief dip during the financial crisis, $CO_2$ emissions are on the rise again. Many climate scientists and policy-makers fear that the world's nations will not act in time to avoid disastrous changes wrought by global warming.

So it was a rare bit of good news when, in February, low-lying, flood-prone Bangladesh, energy powerhouse Canada, eco-conscious Sweden, and climate-change-denier haven the US announced they were joining forces to curtail pollutants that are both exacerbating climate change and adversely affecting human health. Eager to find a strategy to slow global warming, these governments (joined in April by the European Union, Colombia, Japan, Nigeria, and Norway) have initiated a climate change mitigation program that focuses on a less-well-known, but still dangerous, group of pollutants: black carbon (also called black soot), methane, and hydrofluorocarbons.

Together, these pollutants account for about one-third of global warming. Because they are short-lived—black carbon remains in the atmosphere for only days or weeks after it's emitted, whereas carbon dioxide stays in the atmosphere for decades to centuries—curtailing these emissions would have an almost immediate effect. The technological fixes for reducing these pollutants are well tested and, equally important, available and relatively cheap. Experts say that slashing short-term pollutants may prove much easier to accomplish than the large-scale shift away from fossil fuels required for significant $CO_2$ reduction. Success in limiting short-lived pollutants would deliver a much needed psychological boost to the movement to slow global warming and re-energize broader efforts to cut greenhouse gas emissions.

While by no means a panacea for addressing global warming or a substitute for $CO_2$ reduction, the scientific consensus is that reducing black carbon and other short-lived climate pollutants is vital to combating climate change.

"This project holds a lot of promise, especially in the context of our larger battle against climate change," Secretary of State Hillary Rodham Clinton said when she announced the coalition. "Now we know, of course, that this effort is not the answer to the climate crisis. There is no way to effectively address climate change without reducing carbon dioxide, the most dangerous, prevalent, and persistent greenhouse gas. It stays in the atmosphere for hundreds of years. So this coalition is intended to complement—not supplant—the other actions we are, and must be, taking."

To be effective, the coalition needs to get moving immediately. Climate science shows that there's a narrow window in which to act. Scientists estimate that by tackling short-term greenhouse gas pollutants, the currently projected global temperature rise could be reduced by nearly 0.5°C (or 2°F) by 2050. This might not sound like much, but it represents about one-fourth of the temperature rise climate scientists say must be averted to prevent long-term irreversible climate change. And given the international failure to address $CO_2$ emissions, it's likely the best way to avoid some of the most dire climate dislocations.

Black carbon is formed when fossil fuels, biofuels, or biomass—wood, vegetation or dung, for example—is burned but not fully combusted. Its sources include diesel engines—in motor vehicles, boats, trains, and also non-vehicle engines—as well as generators, agricultural burning, wildfires, inefficient cookstoves, and the kind of brick kilns that are common in South Asia. It is also emitted when excess gas from oil rigs and refinery wellheads is burned, or "flared." This type of industrial pollution has created chronic environmental and health problems in regions where refineries are clustered, such as along the US Gulf Coast and Africa's Niger Delta. Picture the heavy black smoke released when a diesel engine starts up, the black plume at the end of a flaring industrial smokestack, or the curls that come off of wood smoke.

Black carbon, simply because it is black, absorbs more light than any other kind of particulate matter. When it settles on ice and snow it absorbs sunlight and makes them melt faster. In Arctic regions, these effects contribute to accelerated snow- and ice-melt and result in greater extents of open polar water that, in turn, allow for increased heat exchange that raises surface and air temperatures. The world's Arctic glaciers are particularly vulnerable to the impacts of black carbon, as are glaciers in high-altitude regions such as the Himalayas. It's estimated that black carbon emissions contribute about half the current warming effects in glaciated regions. This warming speeds spring snowmelt and thus affects when and how much water is available throughout the year for drinking, agriculture, and for natural vegetation. Because glaciers play such a pivotal role in global and regional climate patterns, slowing their melting will help mitigate climate change impacts over the short- and long-term.

"Aggressively reducing black carbon over the next 30 years could reduce currently projected warming in the Arctic by two-thirds," says Erika Rosenthal, staff attorney with the environmental nonprofit Earthjustice and contributing author to the United Nations Environment Program's black carbon assessment.

Black carbon's presence in the atmosphere can also affect cloud formation and lead to disrupted patterns of precipitation. While not directly linked to temperature change, such altered precipitation can have significant impacts on overall climate patterns. Scientists have already begun to link such disruptions—severe storms, droughts, and heat waves and their health effects—to climate change.

Black carbon also contributes to air pollution that has well-known impacts on respiratory and cardiovascular health. Black carbon is a component of the fine particulate pollution known as PM 2.5 (particulate matter of 2.5 microns or smaller). Exposure to particulate pollution contributes to some 2 million premature deaths worldwide each year. Another short-lived climate pollutant, methane, is a precursor to tropospheric ozone that also contributes to poor air quality. (The troposphere is the lowest region of the atmosphere extending from the earth's surface upwards for 6 to 10 kilometers.) The public health benefits of reducing these emissions will be substantial in virtually every region of the world, from heavily urbanized and industrial areas to remote and rural communities.

The good news is that measures to curtail short-lived climate pollutants, particularly black carbon, are readily available, relatively affordable, and don't entail major changes in technology or energy-use habits. Diesel engine retrofits, new engine and fuel emissions requirements, alternate-transportation solutions, and clean-burning cookstoves are some of the proven technologies that could reduce short-term pollutants. For reasons coincidental to climate change mitigation, Rosenthal notes, "efficiency, cost savings, public health benefits" are concurrent goals.

Durwood Zaelke, president of the Washington, DC–based non-profit Institute for Governance and Sustainable Development, believes the new international initiative to curtail these pollutants has enormous potential. Because the results are, at least in theory, so readily achievable, the effort has the potential to "change the psychology" of the climate change movement and to take it "from denial to despair to optimism," he said after the coalition's first meeting in Stockholm. "That these governments have taken this step and so publicly is a major milestone," Rosenthal adds.

Yet there are considerable challenges involved. UNEP experts point out that because the sources of black carbon tend to be so widely distributed and numerous—lots and lots of relatively small-scale sources like individual diesel engines, cookstoves, brick kilns, as well as large-scale sources like agriculture—implementing reductions will require concerted efforts. Luckily, the coalition will not have to start from scratch. It can build on ongoing international clean air and global warming control efforts such as the Montreal Protocol, Global Methane Initiative, Arctic Council, and Global Alliance for Clean Cookstoves. Financial commitments, including a UNEP-managed trust fund and $12 billion in support from the World Bank, are already in place.

Another challenge is quantifying results. That's because emissions sources are varied and each pollutant gets measured differently. The best way to assess progress in reducing black carbon is by looking at specific emission sources, explains Earth justice attorney Rosenthal. For example, one can count how many diesel vehicles have been retrofitted or how many clean-burning cookstoves have replaced inefficient ones, and estimate reductions that way.

It helps to have a picture of the global landscape of such emissions. While biomass burning—including wildfires—contributes about one-third of all black carbon emissions worldwide, contributions from other sources differ in ways that reflect regional variations in economic development. According to the US Environmental Protection Agency's Report to Congress on Black Carbon, released in March 2012, black carbon emissions have been decreasing in North America and Europe over the past century but increasing elsewhere, particularly in developing economies in Asia and Latin America. The US now produces only about 8 percent of the world's black carbon emissions; developing countries, about 75 percent.

The most striking difference in emissions sources is that of residential cookstoves. Globally, about one-quarter of black carbon emissions come from residential sources. Only a small fraction of North Americans depend on biomass-burning stoves. But about 3 billion people worldwide cook and heat their homes with stoves that burn wood, dung, other biomass, or coal. This raises the obvious issue of whether targeting cookstoves for emissions reductions means burdening the less wealthy and pushing the responsibility from those responsible for high per-capita emissions—American SUV drivers, for example—to those already at the low end of the income scale.

But the recognition of how critical these measures are to slowing global warming—combined with the important health benefits they can have—has won over those who were initially skeptical. Key to this, Rosenthal explains, was the decision to separate black carbon reduction efforts from the geopolitics of $CO_2$ reduction and international agreements like the Kyoto Protocol. "Coalition members are very sensitive to these questions," Zaelke says. "They will not advance an agenda that asks their people to do something they cannot do."

Urgency is also a compelling factor. To be effective, short-lived climate pollutants need to be reduced before the world reaches what scientists call "peak warming"—the threshold beyond which climate conditions could become severely detrimental. A January 2012 paper by Drew Shindell—climate scientist at the Goddard NASA Institute for Space Studies—and colleagues, published in *Science*, makes this abundantly clear. To illustrate the impact of these reductions, the paper identifies and maps the effects of 14 different measures that would reduce about 90 percent of current black carbon and methane emissions. These include eliminating high-emission diesel engine vehicles; banning agricultural waste-burning; and replacing inefficient stoves, brick kilns, and coke ovens with clean-burning models.

*It's estimated that black carbon emissions contribute about half the current warming effects in glaciated regions.*

A number of such programs are already underway. Efficient cookstove technology is being supported by the Global Alliance for Clean Cookstoves with funding from the US Department of Energy. In the US, new engine standards, particulate filters, "ultra-low" sulfur diesel fuel, and engine retrofit programs are expected to reduce diesel black carbon emissions by 86 percent by 2030.

But Shindell and colleagues are careful to note that, while the public health and environmental benefits of reducing these short-term climate pollutants can be significant, over the long-term global warming mitigation will depend primarily on reducing $CO_2$ emissions.

"We need to do more on the $CO_2$ side," Zaelke says. "But it's hard to do more when you don't have options. Acting on black carbon and other short-lived climate pollutants will help buy some time. This is something we can do now while we have options."

"Given the already disrupted state of the climate, there's not an either-or here," says Bill McKibben, author and co-founder of 350.org, the international climate campaign. "We've got to do it all. If this is a substitute for reducing carbon, that will be sad. If it's a supplement, it will be a real help."

# Monitoring EU Emerging Infectious Disease Risk Due to Climate Change

By Elisabet Lindgren, Yvonne Andersson, Jonathan E. Suk,
Bertrand Sudre, and Jan C. Semenza
*Science Magazine,* April 1, 2012

In recent years, we have seen transmission of traditionally "tropical" diseases in continental Europe: chikungunya fever (CF) in Italy in 2007, large outbreaks of West Nile fever in Greece and Romania in 2010, and the first local transmission of dengue fever in France and Croatia in 2010.[1-3] These events support the notion that Europe is a potential "hot spot" for emerging and re-emerging infectious diseases (EIDs).[4] Major EID drivers that could threaten control efforts in Europe include globalization and environmental change (including climate change, travel, migration, and global trade); social and demographic drivers (including population aging, social inequality, and life-styles); and public health system drivers (including antimicrobial resistance, health care capacity, animal health, and food safety).[5, 6] Climate change is expected to aggravate existing local vulnerabilities by interacting with a complex web of these drivers.[6] For example, increases in global trade and travel, in combination with climate change, are foreseen to facilitate the arrival, establishment, and dispersal of new pathogens, disease vectors, and reservoir species.

Complexity is not an excuse for inaction. An effective public health response to climate change hinges on surveillance of endemic and emerging diseases.[7] Infectious disease surveillance at the European level is regulated by the European Parliament and Council.[8] In 2009, the European Commission outlined actions needed to strengthen the European Union's (EU) resilience to the impacts of changing climate,[9] specifically on the surveillance of health effects such as infectious diseases. The European Centre for Disease Prevention and Control (ECDC) was charged with guiding this process, which is summarized below. It included extensive literature reviews, evaluation of current surveillance systems in Europe, weighted risk analysis of different diseases, and expert consultation with EU member state representatives.[10, 11]

## Novel Approaches to Risk Assessment and Surveillance Are Needed

Currently, risk analyses tend to focus on identifying diseases that are climate-sensitive. Evidence-based risk assessments with clearly defined uncertainties and underlying assumptions should be routinely undertaken to identify disease risks from climate change and to prioritize possible changes in surveillance. However, a comprehensive climate change assessment needs to be expanded to account for societal aspects so as to fully reflect the impact of climate change. The results from such combined, weighted analyses can be used to evaluate whether surveillance ought to be implemented or modified.

Current EU-wide surveillance is either indicator-based (annual country-level reporting of confirmed human cases) or event-based (detection of individual disease outbreaks through epidemic intelligence). Traditional surveillance often does not suffice for early detection of EIDs. EU member states currently report confirmed human cases of notifiable diseases to ECDC, animal cases of zoonotic diseases to other EU registries, and food-borne and zoonotic outbreaks to the European Food Safety Authority. Reporting of bathing and drinking water quality data are governed by other EU directives, and vector surveillance is voluntary. Early detection of EIDs related to climate change calls for fine-tuning of current surveillance approaches; for example, increased cross-sectoral interactions and sentinel surveillance are warranted.

## Climate Change and Infectious Diseases

Notable changes in annual average temperature and mean precipitation are predicted for Europe, with disproportionately warmer winters in the north and warmer summers in the south.[12] The impacts of climate change on infectious diseases are complex and multifaceted.[11, 13] Changes in precipitation amounts and patterns can bring about increases in water flows and floods, leading to contamination of drinking, recreational, or irrigation water and increased risk of outbreaks of cryptosporidiosis and vero toxin–producing *Escherichia coli* (VTEC) infections, for example. Higher water temperatures increase the growth rate of certain pathogens, such as *Vibrio* species that can cause food-borne outbreaks (seafood) or, on rare occasions, lead to severe necrotic ulcers, septicemia, and death in susceptible persons with wounds bathing in contaminated waters. Elevated air temperatures could negatively affect the quality of foodstuff during transport, storage, and food handling. Increasing temperatures typically shorten arthropod life cycles and the extrinsic incubation periods of vector-borne pathogens, potentially leading to larger vector populations and enhanced transmission risks. Long-term seasonal changes will affect both vectors and host animals and may locally affect land-use changes and human behavior, with implications for the geographical distribution, seasonal activity, and prevalence of many vector-borne diseases in Europe.[11]

## Weighted Risk Analysis

EU member states are currently required to report 46 infectious diseases and 7 additional diseases if they cause hemorrhagic symptoms. In addition, nosocomial infections (i.e., hospital-acquired) and antimicrobial resistance are also mandatory to report. Of the total reportable diseases, 26 were found to be either directly or indirectly affected by climate change, on the basis of a systematic literature review and expert judgment. Of the nonnotifiable diseases, five climate-sensitive diseases were considered to be EIDs due to climate change.[10, 11]

Based on the literature and expert judgments,[10, 11] the strength of association with climate change in Europe was categorized as low, medium, or high. Yet even a strong link with climate change did not necessarily imply the need for surveillance; prevalence, severity, and secondary complications (including human and financial costs for society) were important considerations in this weighted analysis. By combining these parameters, the importance of disease for society was ranked as low, medium, or high. These values were then plotted on a matrix to reflect the joint importance of climate change and severity of a specific disease in Europe.

Several of the climate-sensitive reportable diseases are found in the medium category. Most of these are already under adequate surveillance, even if climate change further increases disease risk, but seven could justify changes in the surveillance system. Hemorrhagic cases of dengue fever (DF) and Rift Valley fever (RVF) are currently reported to ECDC but are all imported cases. However, increased risk of autochthonous (indigenous, not imported) transmission demands heightened surveillance.

## Surveillance of EID Threats from Climate Change

The recent emergence of DF and CF in Europe has been traced to environmental conditions conducive to high vector population densities of the invasive Asian tiger mosquito, *Aedes albopictus*, and permissive temperatures for pathogen replication within the vector.[1] Surveillance of DF and CF will require a combination of vector surveillance—to detect new risk areas and human case reporting—with a focus on the warmer parts of the year to detect autochthonous cases.

Monitoring established vectors close to their altitude and latitude distribution limits will help to delineate expanding and contracting areas of risk. Human case reporting of Lyme borreliosis could potentially be linked to sentinel monitoring of infected ticks in selected areas, a practice already implemented in the Czech Republic and planned for Denmark. Using the disease-specific skin rash, erythema migrans, as the case definition of Lyme borreliosis would pick up shifting distribution and risks, rather than the rare laboratory-confirmed Lyme neuroborreliosis. For the detection of possible new risk areas of leishmaniasis, surveillance of the sandfly vectors could be coupled with monitoring of infected pet dogs in the EU. Tick-borne encephalitis (TBE) is best surveyed through reporting of serologically confirmed human cases.

Establishing surveillance for the introduction of new vector species into the EU could contribute substantially to infectious disease control, particularly when linked with surveillance of imported human and/or animal cases. Enhanced collaboration between the veterinary surveillance and public health sector will advance preparedness and response if pathogens and vectors become prevalent in the region and pose a threat to humans. Lessons learned from pandemic avian flu preparedness could be applied to other EIDs. Collaboration between the human and veterinary sectors was considerably improved through such practices as regular meetings, routine sharing of epidemiologic and laboratory data, preparation of linked response plans for human and veterinary health, and coordinated outbreak investigations.

> *Changes in precipitation amounts and patterns can bring about increases in water flows and floods, leading to contamination of drinking, recreational, or irrigation water. . . .*

Important food-borne and waterborne diseases, in particular the zoonoses, are already part of surveillance collaborations in Europe. However, new collaborations will be needed. Surveillance of *Vibrio* spp. outbreaks due to seafood consumption could be enhanced by initiating collaborations between human case surveillance and laboratory monitoring of bivalve shellfish contamination.

These challenges call for the development of improved surveillance by monitoring environmental precursors of disease. ECDC is developing the European Environment and Epidemiology (E3) Network, designed to link environmental with epidemiological data for early detection of and rapid response to shifting infectious disease burden.[11] ECDC is also developing VBORNET, a surveillance tool used for monitoring the distribution of invasive disease vectors within the EU.

## Conclusions

Adjustments to existing surveillance practices in the EU, as outlined above, will enhance preparedness and facilitate the public health response to EIDs and thereby help contain human and economic costs.[14] These benefits need to be balanced against the additional costs of program implementation; however, cost-benefit analyses tend to be hampered by the inherent difficulty in attributing EID impacts to specific climate change events. Nevertheless, these policy-driven adaptation measures should facilitate early detection of EIDs, regardless of the underlying macro-level drivers.

## References and Notes

1. G. Rezza et al, *Lancet* 370, 1840 (2007).
2. A. Papa et al, *Clin. Microbiol. Infect.* 17, 1176 (2011).
3. E. A. Gould et al, *Clin. Microbiol. Infect.* 16, 1702 (2010).
4. K. E. Jones et al., *Nature* 451, 990 (2008).
5. R. A. Weiss, A. J. McMichael, *Nat. Med.* 10 (suppl.), S70 (2004).
6. J. E. Suk, J. C. Semenza, *Am. J. Public Health* 101, 2068 (2011).
7. D. A. King et al., *Science* 313, 1392 (2006).
8. Commission of the European Communities, Decision no. 2119/98/EC of the European Parliament and of the Council (1999).
9. Commission of the European Communities, Adapting to Climate Change: Towards a European Framework for Action COM (2009) 147 final (Brussels, 2009).

10. J. C. Semenza, J. E. Suk, V. Estevez, K. L. Ebi, E. Lindgren, *Environ. Health Perspect.* 120, 385(2012).
11. J. C. Semenza, B. Menne, *Lancet Infect. Dis.* 9, 365 (2009).
12. F. Giorgi, X. Bi, J. Pal, *Clim. Dyn.* 23, 839 (2004).
13. S. E. Randolph, D. J. Rogers, *Nat. Rev. Microbiol.* 8, 361 (2010).
14. Marsh Inc., *The Economic and Social Impact of Emerging Infectious Disease* (New York: Marsh Inc., 2008).

# Bibliography

❖

Boykoff, Maxwell T., and Susanne C. Moser. *Successful Adaptation in Climate Change*. Abingdon: Routledge, 2013. Print.

Bulkeley, Harriet. *Cities and Climate Change*. Abingdon: Routledge, 2013. Print.

Cadman, Timothy. *Climate Change and Global Policy Regimes: Towards Institutional Legitimacy*. New York: Macmillan, 2013. Print.

Cossia, Juliann M. *Global Warming in the 21st Century*. New York: Nova, 2011. Print.

Dutch, Steven I. *Encyclopedia of Global Warming*. Pasadena: Salem, 2010. Print.

El-Nemr, Ahmed. *Environmental Pollution and Its Relation to Climate Change*. Hauppauge: Nova, 2011. Print.

Fouquet, Roger. *Handbook on Energy and Climate Change*. Cheltenham: Elgar, 2013. Print.

Grover, Velma I. *Impact of Climate Change on Water and Health*. Boca Raton: Taylor & Francis, 2013. Print.

Henderson-Sellers, A., and K. McGuffie. *The Future of the World's Climate*. Amsterdam: Elsevier, 2012. Print.

Letcher, T. M. *Climate Change: Observed Impacts on Planet Earth*. Amsterdam: Elsevier, 2009. Print.

Pielke, Roger A. *Climate Vulnerability: Understanding and Addressing Threats to Essential Resources*. Amsterdam: Academic P, 2013. Print.

Pryor, S. C. *Climate Change in the Midwest: Impacts, Risks, Vulnerability, and Adaptation*. Bloomington: Indiana UP, 2013. Print.

Richardson, Katherine. *Climate Change: Global Risks, Challenges and Decisions*. Cambridge: Cambridge UP, 2011. Print.

Robarts, R. D., Michio Kumagai, and Charles Remington Goldman. *Climatic Change and Global Warming of Inland Waters: Impacts and Mitigation for Ecosystems and Societies*. Chichester: Wiley & Sons, 2013. Print.

Rosenzweig, Cynthia, and Daniel Hillel. *Handbook of Climate Change and Agroecosystems: Global and Regional Aspects and Implications*. London: Imperial College P, 2013. Print.

Salih, Mohamed Abdel Rahim M. *Local Climate Change and Society*. Abingdon: Routledge, 2013. Print.

Steffen, W., Sanderson, A., Tyson, P. D., et al. *Global Change and the Earth System: A Planet under Pressure*. Heidelberg: Verlag, 2004. Print.

Sundaresan, J. *Climate Change and Island and Coastal Vulnerability*. Dordrecht: Springer, 2013. Print.

Weart, S. R. *The Discovery of Global Warming*. Cambridge: Harvard UP, 2008. Print.

# Websites

❖

### Center for Climate and Energy Solutions
http://www.c2es.org/

Originating as the Pew Center on Climate Change, C2ES serves as an independent research program on the policy and science of climate change. The organization is recognized as a leading think tank on the subject.

### The Climate Institute
http://www.climate.org/

The mission of the organization is to provide practical solutions to climate change mitigation by creating partnerships among policymakers, scientists, and public and environmental institutions.

### EPA: United States Environmental Protection Agency
http://www.epa.gov/climatechange/

The EPA is the government agency responsible for US policy in the protection of human health and the environment. The website provides information on the president's policy on climate change and recommended actions.

### National Center for Atmospheric Research
http://ncar.ucar.edu/

NCAR is a research and development center devoted to service, research, and education in the atmospheric and related sciences.

### NOAA: National Climatic Data Center
http://www.ncdc.noaa.gov/

The organization maintains the world's largest climate data archive in support of every sector of the United States economy and users worldwide.

### The United Nations World Meteorological Organization
http://www.wmo.int/pages/index_en.html

A specialized agency within the United Nations, the WMO serves as the voice of the UN on national and global policy on climate change.

### US Global Climate Change Research Program
http://nca2009.globalchange.gov/

The website provides information on the 2009 National Climate Assessment report on the impact of climate change in the United States.

# Index

❖